Next-Gen Manufacturing 2019

Topical Conference at the 2019 AIChE Annual Meeting

Orlando, Florida, USA
10 - 15 November 2019

ISBN: 978-1-7138-0555-7

Printed from e-media with permission by:

Curran Associates, Inc.
57 Morehouse Lane
Red Hook, NY 12571

Some format issues inherent in the e-media version may also appear in this print version.

Printed with permission by Curran Associates, Inc. (2020)

For permission requests, please contact AIChE
at the address below.

AIChE
120 Wall Street, FL 23
New York, NY 10005-4020

Phone: (800) 242-4363
Fax: (203) 775-5177

www.aiche.org

Additional copies of this publication are available from:

Curran Associates, Inc.
57 Morehouse Lane
Red Hook, NY 12571 USA
Phone: 845-758-0400
Fax: 845-758-2633
Email: curran@proceedings.com
Web: www.proceedings.com

TABLE OF CONTENTS

571591 Demystifying Mass Transfer in Electrolyzers through Rapid Prototyping with 3D Printed Parallel Plate Cells

As a result of the global energy transition, a large percentage of the world's power supply will become electric. For electrochemical engineers this transition generates a sea of opportunities. After all, it is electrolyzers that use electricity to produce some of life's essential chemicals. However, fully embracing these opportunities in a sustainable way is challenging: Future electrolyzers must be safer, cleaner, and far more flexible, while simultaneously increasing productivity and cost-effectiveness. Essentially, they will have to achieve more with less.

One key parameter in the performance of electrolyzers is the rate of mass transfer. Most often, electrochemical reactions are mass transfer limited due to the fast kinetics that depend exponentially on the applied overpotential. Numerous correlations in literature exist to describe the rate of mass transfer in a specific reactor.[1] Though, their predictions can vary up to almost an order of magnitude (see figure 1). The reason for this is the large geometric variety between the reactors that were measured. The design of the inlets in particular is a crucial contributor to the rate of mass transfer. [2] Moreover, only a few lab-scale electrolyzers are commercially available, often with sub-optimal designs, which limits further research. As a consequence, it is difficult to establish a generalized description of the mass transfer performance of parallel plate electrolyzers. Furthermore, most existing correlations only describe the average rates of mass transfer and therefore do not consider any local variation despite the importance thereof.[3]

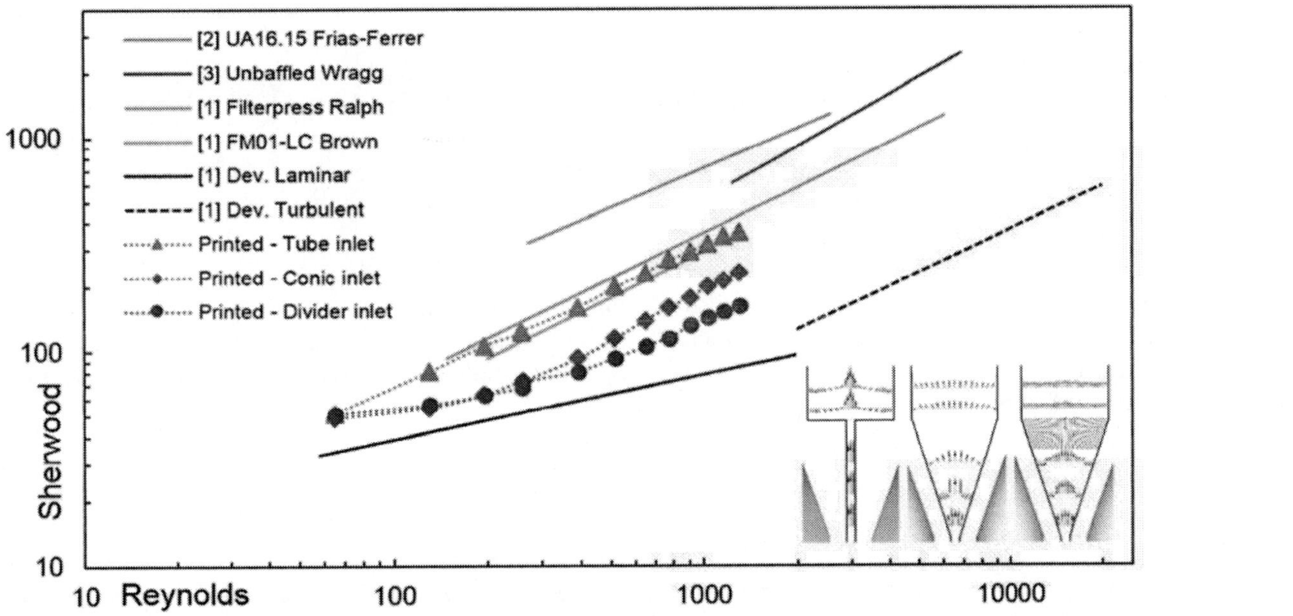

Figure 1: Mass transfer performance in terms of Sherwood vs. Reynolds. Up to an order of magnitude of difference in performance is visible between reactors. The red triangle, square and circle markers denote a selection of inlet manifolds tested with the 3D printed prototype. Small insert: sketches of the inlet manifolds, from left to right: tube inlet, conic inlet, divider-type inlet.

The goal of our study has been to demystify the fundamentals of mass transfer in parallel plate electrolyzers. As this would have required testing a large geometric variety of cells, traditional manufacturing techniques were impractical due to their lead time and cost. Therefore, we have used 3D printing to rapidly construct many different reactor prototypes. Moreover, each prototype could be finetuned to the exact specifications each measurement required. This way, each geometric parameter could be varied and investigated individually.

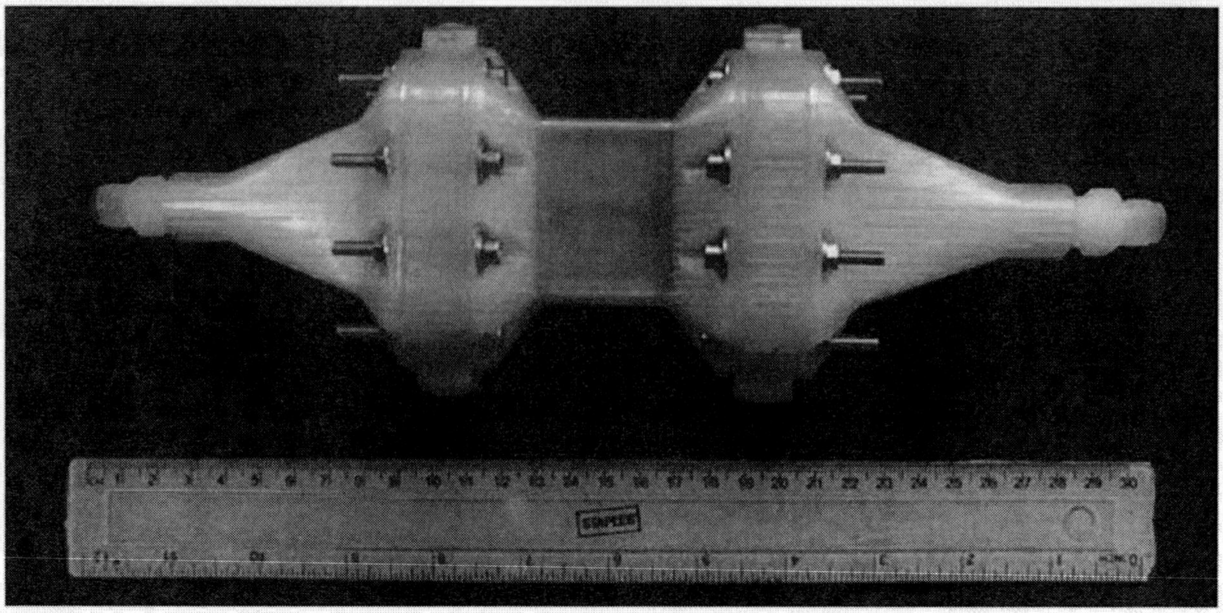

Figure 2: 3D printed electrolyzer prototype for mass transfer studies. The electrode area is 40 cm^2.

Nearly the entire reactor can be 3D printed in less than a workweek's time (see figure 2). Inlet manifolds, the channel and the rubber sealings could all be printed using commercially available equipment. Furthermore, by using 32 independently operated electrode segments we were able to measure the mass transfer performance on a local scale. Sherwood – Reynolds correlations were established for designs with different shapes of turbulence promoters, 3D electrodes and inlet manifolds. The effect of an increased or decreased hydrodynamic entrance length was also investigated.

In the end these prototypes lead to enhanced understanding of electrochemical reactors, allowing us to intensify processes such as alkaline water electrolysis. As a proof of concept, we are developing a 3D printed lab-scale electrolyzer for alkaline water splitting which can be used to test the effect of geometry on the production of hydrogen directly.

References

[1] T.R. Ralph, M.L. Hitchman, J.P. Millington, F.C. Walsh, *Mass Transport in an electrochemical laboratory filterpress reactor and its enhancement by turbulence promoters,* Electrochem. Acta, **1994,** pp. 591-603.

[2] A. Frías-Ferrer, J. González-García, V.Sáez, C. Ponce de León, F.C. Walsh, *The effects of manifold flow on mass transport in electrochemical filter-press reactors,* AIChE J., **2008,** pp. 811-823.

[3] A.A. Wragg, A.A. Leontaritis, *Local mass transfer and current distribution in baffled and unbaffled parallel plate electrochemical reactors,* Chem. Eng. J., **1997,** pp. 1-10.

574364 Additive Manufacturing of Catalyst Support Structures with in Operando Adjustable Mass Transport and Flow Characteristics

In conventional randomly packed bed reactors, the flow and mass transport characteristics are predefined by the particle diameter and the tube diameter. Structured reactors offer the opportunity to tailor mass transport and flow field according to the needs of the reaction system. In this regard, additively manufactured so-called Periodic Open Cellular Structures (POCS) as catalyst supports have shown their advantages over conventional randomly packed bed reactors in terms of heat management and pressure drop [1, 2]. However, since the internal flow field, pressure drop and therefore the resulting mass transport characteristics are mainly influenced by the geometrical properties of the unit cell, for POCS the resulting transport characteristics of the structure is already determined and fixed by the choice and design of the unit cell. To achieve more flexibility, we have recently proposed a new type of so-called interpenetrating POCS (interPOCS) which are formed by inserting a diamond unit cell based structure in the cavities of a second diamond cell structure. These interPOCS allow for in operando flow field and mass transport adjustments by shifting the (relative) position of the loose structure [3].

With detailed computational fluid dynamics (CFD) based simulations with an additional in-house particle-tracking algorithm implementation we systematically investigated the flow field and mass transport characteristics within the interPOCS in dependency of the relative positioning of the loose structure. The highly adjustable structure allows for a broad variation of mass transport characteristics in a catalytic reactor without changing the system's periphery. This functionality enables mass transport optimization for a given system and represents a completely new application of additively manufactured catalyst support structures for highly flexible in operando tuning of flow field and mass transport characteristics.

References

[1] C. Busse, H. Freund, and W. Schwieger (2018). *Chem. Eng. Process.* 124, 199-214.

[2] A. Inayat, M. Klumpp, M. Lämmermann, H. Freund and W. Schwieger (2016). *Chem. Eng. J.* 287, 704-719.

[3] G. Do, T. Stiegler, M. Fiegl, L. Adler, C. Körner, A. Bösmann, H. Freund, W. Schwieger and P. Wasserscheid (2017). *Ind. Eng. Chem. Res.* 45, 13402-13410.

561588 Role of 3D Printed and LED-Driven Photostation in Photocatalytic and Photo-Fenton Activity of Iron Oxide Doped Graphitic Carbon Nitride

The degree of photocatalysis or any photolysis reaction depends on the intensity of light. In most cases, a major portion of light is not used as it does not reach the reaction zone thereby wasting a significant portion of light and energy, as well. In order to handle this situation, we have designed and 3D printed a photostation with complex parabolic reflective surface, so that all the incident light beams can reach the reaction zone. The photostation is fitted with two chips-on-board (COB) light emitting diodes (LEDs) reaching the total (visible) light intensity within the reaction zone over 290,000 lux (For comparison, it is commonly reported that the sun emits anywhere from 32,000 to 100,000 lux onto the Earth). Light simulation analysis confirmed that all incident light beams can reach the reaction zone of photostation. In order to test the performance of this photostation, we have used 0.5 to 2 wt.% iron oxide doped graphitic carbon nitride (g-C_3N_4) to degrade a model water pollutant, dicamba (3,6-Dichloro-2-methoxybenzoic acid), which is common herbicide, in presence of traces of hydrogen peroxide and visible light. It has been confirmed experimentally that the rate of degradation of dicamba is directly proportional to the light intensity and geometry of photostation thereby confirming the positive role of photostation design. The rapid degradation of dicamba under this study can be attributed to the four distinct phenomena of photocatalytic activity of carbon nitride, quenching of electron/hole pairs and generation of additional reactive hydroxyl radicals by hydrogen peroxide, Fenton and photo-Fenton activity of iron oxide component of carbon nitride in presence of hydrogen peroxide and photocatalytic activity of iron oxide alone in conjuncture with carbon nitride. Liquid Chromatography Mass spectrometry (LCMS) analysis of the degradation products showed loss of chlorine from the aromatic ring and evidence of free radical addition reactions in the course of photocatalysis.

557053 Solution-Based 3D Printing of Hierarchical Microporous Polymers for Energy-Efficient Adsorption

Current 3D printing methods have strong limitations in the classes of polymers that can be printed. This means that many polymers of technological interest cannot currently be 3D printed. Here, we demonstrate a generalizable method for 3D printing of viscoelastic polymer solutions and apply it to both porous (a polymer of intrinsic microporosity, PIM-1) and non-porous (e.g., cellulose acetate, Matrimid) polymers. Successful ternary ink formulations that result in the creation of three-dimensional structures with hierarchical porosity require balancing of solution thermodynamics (phase separation), mass transfer (solvent evaporation), and rheology. To demonstrate the utility of this approach, a microporous polymer (PIM-1) that cannot be printed with current 3D printing technologies is 3D printed into a high-efficiency respirator exhibiting hierarchical porosity ranging from sub-nanometer to millimeter pores. These short mass transfer contactors (1.27 cm) were challenged with a flow of toluene vapor (10000 ppm) in N_2 gas. Owing to the precisely designed gas distribution channels, the 3D printed respirator were found to create a pure (< 1 ppm toluene) N_2 stream for 1.7 hours, 6 times longer than traditional structures fabricated from the same PIM-1 material, and more than 4,000 times the residence time of gas in the bed. This solution-based additive manufacturing approach significantly extends the material spectrum of 3D printing technologies. By 3D printing adsorptive materials into precisely designed mass transfer contactor, the performance, energy efficiency, and material efficiency can be significantly improved.

559790 Titanium/Titania Electrocatalyst for Hydrogen Evolution Synthesized Via a Combination of 3D Printing and Anodization

A titanium core/titania coating, hierarchically structured catalyst was fabricated via combining 3D printing and in-situ anodization. 3D printing enabled control of the macro-porosity of the titanium core. Anodization was used in order to grow highly-ordered TiO_2 nanotubes (NTs) on the metal core. The dimensions of the TiO_2 nanotubes were controlled via tuning of anodization parameters. The combination of both fabrication techniques allows for hierarchical control of the catalyst structure.

The performance of the reported 3D Ti/TiO_2 electrocatalyst was tested for hydrogen-evolution reaction (HER) via water electrolysis. Titanium dioxide is a known photo-electro-catalysts for water splitting, but competitive electrocatalytic activity for HER in the absence of light has not been reported. Onset potentials for HER as low as -45 mV vs. RHE were obtained for our electrodes in acid medium, making them competitive with current state-of-the-art electrodes aiming to replace platinum. Tafel slopes between 40 and 52 mV dec^{-1} point towards a Volmer-Heyrovsky mechanism, with the Heyrovsky reaction as possible controlling step.

The proposed electrocatalysts feature the following advantages: (1) the fabrication method for the 3D Ti/TiO_2 catalyst is simple, scalable and can be readily incorporated as printed reactor components into different technologies; (2) the 3D Ti/TiO_2 catalyst features a flow-through architecture that not only enables more efficient utilization of the surface area of the catalyst but that could enable large-scale, continuous-flow hydrogen-production reactor technologies; and (3) minimization of electrical resistance between active surface area (TiO_2 NTs) and current collector.

556480 Advanced Manufacturing Enabling Process Intensification for Catalytic Processes

Advanced manufacturing is a transformative and disruptive approach to industrial production that enables the rapid synthesis of virtually any geometry composed of a wide range of materials, such as polymers, ceramics, and metals. Process intensification is a set of novel approaches to process engineering and equipment design that leads to substantially smaller, more selective, and energy-efficient processing plants. The integration of advanced manufacturing with process intensification accelerates process development and deployment through rapid prototyping and enables new hardware that overcomes transport limitations. Advanced manufacturing (ex. 3D printing or additive manufacturing) is the physical backbone of Industry 4.0 that can take advantage of the conclusions of big data and machine learning by allowing design freedom. This expansion of the material design window can be utilized to reduce GHG intensity for novel or existing processes.

Numerous conversion technologies in the energy industry suffer from rapid coke build-up and subsequent deactivation of a catalytic material that is exposed to a carburizing environment. The time scale of deactivation often determines the reactor technology and process configuration employed. Commercial examples of deactivating catalysts range from seconds in a fluid bed catalytic cracking unit to minutes for moving bed propane dehydrogenation and days for continuous catalytic reformers or even years for steam methane reformers. A common step for these processes is the regeneration of the catalytic material via combustion of the coke. However, the cyclic process of coking in a carburizing environment followed by combustion in an oxidizing environment can often be detrimental to catalyst performance.

This work will discuss how advanced manufacturing methods can enable construction of materials capable of rapidly switching between two different environments with enhanced catalytic activity. Such materials capable of withstanding cyclic carburizing and oxidizing steps can be efficiently integrated into novel reactor concepts. The conversion of light gas is an excellent probe reaction for elucidating this behavior. In an effort to search for robust materials, honeycomb structures composed of metal were fabricated by direct metal laser sintering (DMLS), a 3D printing technique that involves the layer-wise successive deposition of micron-sized powder followed by laser induced sintering. The cyclic operation that alternates between carburizing and oxidizing environments leads to compositional modification at the surface of the monolith channels that imparts in-situ catalytic activity.

565296 Collaborative Simulation Development Accelerated By Cloud Based Computing and Software As a Service Model

Simulations are increasingly used in pharmaceutical development to deliver medicines to patients more quickly; more efficiently; and with better designs, safety, and effect. Many of the risks in product design, including scale up can be mitigated by modeling and simulation. Improved product understanding and design space delineation, as well as the possibility for control of complex continuous manufacturing equipment are possible.

These pharmaceutical development oriented simulations need high performance computing resources as well as a variety of software to model everything from molecular dynamics to entire processes trains for the effects on the pharmaceutical product at various scales of scrutiny. The demand curve for these resources has many peaks and can shift in a time scale much faster than a typical procurement process.

Both on demand cloud based computing capability and software as a service models have been growing in use. This presentation describes the efforts of the Enabling Technology Consortium to apply these information technology models to pharmaceutical simulations which have special needs for documentation and security. It is expected that the environment will have more benefits as the cloud can be configured for collaborative work among companies in the non-competitive space and work can be made available for collaboration and use by contract service vendors or health authorities.

The expected benefits of this computing environment include economies of scale for both the providers and the consumer, increased speed, and a higher level of value generation for providers by the use of increased resources and for consumers by the information available to accelerate and improve delivery of pharmaceutical products.

563390 Rational Solvent Selection Guided By Machine Learning and Molecular Descriptors in Asymmetric Catalytic Reactions

Rational solvent selection guided by machine learning and molecular descriptors in asymmetric catalytic reactions

Yehia Amar,[a] Artur M. Schweidtmann,[b] Liwei Cao,[a,d] Paul Deutsch[c] and Alexei Lapkin [a,d]

[a] Department of Chemical Engineering and Biotechnology, University of Cambridge, Philippa Fawcett Drive, Cambridge, CB3 0AS, United Kingdom

[b] Aachener Verfahrenstechnik – Process Systems Engineering, RWTH Aachen University, Aachen, Germany

[c] UCB Pharma S.A. Allée de la Recherche, 60 1070 Brussels, Belgium

[d] Cambridge Centre for Advanced Research and Education in Singapore Ltd, 1 Create Way, CREATE Tower #05-05, 138602, Singapore

Email: aal35@cam.ac.uk

Abstract

Rational solvent selection remains a significant challenge in process development, especially within pharmaceutical applications. Therefore, a hybrid mechanistic - machine learning approach, geared towards automated process development work-flow was developed, and successfully applied on a $Rh(CO)_2(acac)$/Josiphos (R_1=cyclohexyl, R_2=4-methoxy,3,5-dimethylphenyl) catalyzed asymmetric hydrogenation of a chiral a-ß unsaturated γ-lactam reaction. The mechanistic part of the model is based on molecular descriptors of physico-chemical properties of solvents, including the reaction-specific descriptors, substrate and hydrogen solubility. A library of 400 solvents was used with 17 molecular descriptors.

The algorithm, which is based on a Gaussian process surrogate model, was trained to learn and optimize for both conversion and diastereomeric excess simultaneously, ultimately identifying better solvents.

In addition to being a powerful design of experiments methodology, the resulting statistical surrogate model is predictive, with a cross-validation correlation coefficient of 0.83. Furthermore, a solvent-mixing strategy based on the black box approach was also investigated. These methods open the door for process chemists to use enhanced process development workflows for optimization and discovery.

571487 Application of Multivariate Statistical Process Control to a Drying Process for Continuous Pharmaceutical Manufacturing

1. Introduction

In the pharmaceutical industry, the method for the improvement of the production efficiency has been discussed and the documents on Quality by Design (QbD) and Process Analytical Technology (PAT) [1-4] were published by Food and Drug Administration (FDA) and International Conference on Harmonisation of Technical Requirements for Registration of Pharmaceuticals for Human Use (ICH). Since then, online process monitoring and control technologies have attracted much attention. These technologies are important to realize the continuous manufacturing, and a new draft guideline that addresses quality aspects in the continuous manufacturing of medicinal products are published by the FDA in 2019 [5].

Data-driven multivariate analysis is an important part of PAT and many research papers were published [6, 7]. Latent variable method such as principal component analysis (PCA) and partial least squares (PLS) are popular, and they are applied to various kinds of processes such as crystallization, mixing, granulation, drying, and coating [6]. In this research, multivariate statistical process control (MSPC) is applied to the drying process. The research on MSPC in the drying process is rarely reported. As far as the authors know, only one paper [8] reported the application results of MSPC to a drying process and A.F. Silva et al [8] concluded that the abnormal conditions were successfully detected by MSPC. However, the rate of false positive is quite high: Q statistics are always over the threshold in some cases with normal operation conditions. This research aims to realize more accurate abnormal detection.

2. Experiments

Eight experiments were conducted under the normal operation conditions, and six experiments were conducted under the abnormal operation conditions. The abnormal operation conditions were caused by changing the flowrate, the temperature, and the humidity of the inlet air. Five variables, i.e. the inlet air flowrate, the inlet air temperature, the outlet air temperature, the product temperature, and the NIR spectrum of the product, were recorded every ten seconds.

3. PCA based MSPC

The data from seven experiments with normal operation conditions were used for model construction. Multiway PCA with one principal component was used since the drying process is batch process, and the model was constructed for every batch time. The maximum values of Hotteling's T^2 and Q statistics for the model construction data are selected as the threshold for the abnormal detection. The model validation was conducted by using the data from one experiment with normal operation condition and six experiments with abnormal operation conditions. The MSPC method could classify the operation condition with the accuracy of more than 95%.

References

[1] ICH, ICH harmonised tripartite guideline - pharmaceutical development Q8 (R2) (2005)

[2] ICH, ICH harmonised tripartite guideline - quality risk management Q9 (2005)

[3] ICH, ICH harmonised tripartite guideline - pharmaceutical quality system Q10 (2008)

[4] FDA, Pharmaceutical cGMPs for the 21st century—a risk-based approach final report (2004)

[5] FDA, Quality considerations for continuous manufacturing guidance for Industry (2019)

[6] T. Rajalahti and O. M. Kvalheim "Multivariate data analysis in pharmaceutics: a tutorial review", International journal of pharmaceutics, 417, (2011) 280-290

[7] G. M. Troup, C. Georgakis, "Process systems engineering tools in the pharmaceutical industry", Computers and chemical engineering, 51 (2013) 157-171

[8] A.F. Silva et al. "Multivariate statistical process control of a continuous pharmaceutical twin-screw granulation and fluid bed drying process", International journal of pharmaceutics, 528, (2017) 242-252

571490 Multivariate Data Analysis to Drive Process Understanding and Optimization

Chemical and bio processes can be sensitive to a wide range of parameters. Significant amounts of data are generated to find certain Critical Quality Attributes (CQAs) not meeting specifications. In such cases, a retrospective analysis helps identify trends and patterns in the data. Multivariate data analysis (MVDA) is a very powerful tool which uses various algorithms to elucidate inter-parameter interactions; as well as correlations with output variables. This presentation involves four case-studies wherein MVDA has been implemented to enhance process understanding.

- Spectroscopic methods are widely used for tracking reactions and crystallizations in pharmaceutical processes. In many cases, entire spectra (rather than a single peak specific to a molecule) need to be tracked to accurately represent the process, thus creating a complex multivariate data-set. In this example, a Partial Least Squares (PLS) regression was performed on Raman spectra to understand thermodynamic dependence of crystal forms over a range of temperatures and solvent compositions.
- A lipid nanoparticle formulation was manufactured in a recent manufacturing campaign, with several process parameters and output variables. Principal Component Regression (PCR) provided correlations between parameters and each variable, while PLS regression provided overall scores and loadings plots to eliminate outliers amongst batches.
- Process parameters for a cell-culture process were investigated to quantify influence on cell growth and product quality. Trajectories were generated using the scores plot to track evolution of the batches over time and projection models showed how failed batches deviated from ideal runs. The correlation loadings plot highlighted the continuous variables which had the strongest influence at each stage of the process.
- Data was generated from a series of product stability experiments for two monoclonal antibodies. Multivariate Linear Regression (MLVR) was used to correlate independent variables to CQAs. Results were verified using a PLS regression and a predictive mechanistic model was created.

In each of the above examples, MVDA was able to extract useful insights and successful recommendations were made for process improvement.

577656 Dissolution Prediction By Process Analytical Technology, Machine Learning and Mathematical Modeling: Toward the Real-Time Release Testing of Pharmaceuticals

Continuous pharmaceutical manufacturing, both at upstream (API production) and downstream (formulation technology) processing, is rapidly gaining interest as a promising way to improve efficiency, reduce operating costs and ensure easier scale-up and shorter time-to-market [1]. However, the risk mitigation of process failure is higher when the unit operations are integrated into one continuous process line as the failure can cumulatively disturb the whole line [2]. To fully exploit the system-wide performance of a continuous manufacturing line, while providing the required product quality, the entire process must be flexibly controlled to compensate the process variations. Consequently, the conversion from batch to continuous manufacturing also entails the necessity of changing the current quality assurance approaches. Addressing these challenges, the Process Analytical Technology (PAT) and Quality-by-Design (QbD) initiatives emerged over the last few years to support the design, analysis and control of critical quality attributes (CQA) and critical process parameters (CPP) of the pharmaceutical processes. These can also lead to the feasibility of real-time release. The real-time release testing (RTRT) methods [3] utilize a combination of measured material attributes and in-process analysis to demonstrate that the product conforms to the quality attributes defined by the regulations, without the need for end-product quality tests.

In this work, the prediction of *in vitro* dissolution properties of pharmaceutical tablets is studied using PAT data, machine learning and mathematical modeling of the continuous process. *In vitro* dissolution testing is one of the most substantial analytical methods in the pharmaceutical industry. It is widely applied during the research and development stage, such as for formulation optimization, prediction of in vivo performance or can be used as a surrogate measurement for bioequivalence. Dissolution measurement also serves as a routine quality control (QC) test during the manufacturing process to assess product performance, stability or batch-to-batch variability [4]. In this study, the dissolution of two model pharmaceutical tablet formulations was studied, an immediate release acetylsalicylic acid formulation as well as caffeine tablets with extended-release dissolution properties. The dissolution curves of the formulations were simultaneously influenced by several critical factors: the concentration of the active pharmaceutical ingredient (API), the concentration of the polyethylene oxide, used to create the extended-release matrix, the tableting compression force and the particle sizes of raw materials. Reflection and transmission NIR and Raman spectra were measured on tablets prepared by following experimental designs as well process data were acquired during the manufacturing.

Several modeling approaches were tested and compared to predict the dissolution of the tablets based on the PAT and process data. Partial Least Square (PLS) regression and feedforward backpropagation Artificial Neural Network (ANN) models were developed for the different spectroscopic measurements individually as well as by combining them together. The decrease of the root mean square errors (RMSE) of the models and the improved prediction of the dissolution curves belonging to independent validation tablets indicated a consistently enhanced performance of the ANN models, which could be explained by their capability of modeling non-linearity between the process parameters and dissolution curves. To move forward the real-time release testing approach, the developed dissolution prediction models could also be fitted into a system-wide process model of a continuous manufacturing line, which describes the variation of the identified critical parameters (*e.g.* API concentration, particle size) throughout the manufacturing process steps. In this way, the effect of mitigation of process parameter variability on the dissolution properties could be studied, which can also serve as a basis of process control.

The results of this work indicate that data-driven models could provide a non-destructive and fast alternative for the current quality control standard, in line with the real-time release initiative. Machine learning also served as a straightforward data fusion method of the PAT data without the need for additional preprocessing steps. The method could significantly increase the efficacy of dissolution testing, advance data processing in the PAT environment and contribute to an enhanced real-time release testing procedure, especially when the dissolution model is incorporated into a system-wide modeling approach.

[1] S.L. Lee, T.F. O'Connor, X. Yang, C.N. Cruz, S. Chatterjee, R.D. Madurawe, C.M.V. Moore, L.X. Yu, J. Woodcock, Modernizing pharmaceutical manufacturing: From batch to continuous production, Journal of Pharmaceutical Innovation, 10 (2015) 191-199

[2] A. Mesbah, J.A. Paulson, R. Lakerveld, R.D. Braatz, Model predictive control of an integrated continuous pharmaceutical manufacturing pilot plant, Organic Process Research & Development, 21 (2017) 844-854.

[3] European Medicines Agency, 2012. Guideline on Real Time Release Testing (formerly Guideline on Parametric Release)

[4] Dressman, Jennifer B., and Johannes Krämer, eds. Pharmaceutical dissolution testing. Boca Raton, FL:: Taylor & Francis, 2005.

556721 An Integrated Approach to Modelling, Data, and Control for Pharmaceutical Manufacturing

An integrated Quality-by-Design approach for pharmaceutical manufacturing is described. Continuous manufacturing is an approach that brings process research and development into closer alignment and execution with manufacturing, by significantly reducing the amount of scale up needed to go from laboratory scale to production scale. Dynamic process and plant models are used for integration across different value streams, from a seamless connection from raw materials through drug substance and drug product. The streamlined process development workflow incorporates best practices from the relevant ICH guidelines [1-4], including the selection of control strategies for each critical quality attribute, and the effective use of modular plant-wide simulation. The main concepts are illustrated for small-molecule pharmaceutical pilot plants and bench-scale biologic drug manufacturing platforms. Additional challenges associated with biologic drugs are discussed. Aided with in-line process analytical technology, automated plant-wide control systems are designed and implemented to satisfy specifications on the critical quality attributes. The presentation draws from Quality by Design strategies that were developed and applied in the pharmaceutical industry from before the ICH guidelines were published [e.g., 5-7], up to recent advances [8-22].

References

1. ICH Q8(R2) Pharmaceutical Development, ICH Harmonised Tripartite Guideline, International Conference on Harmonisation of Technical Requirements for Registration of Pharmaceuticals for Human Use, 2009
2. ICH Q9 Quality Risk Management, ICH Harmonised Tripartite Guideline, International Conference on Harmonisation of Technical Requirements for Registration of Pharmaceuticals for Human Use, 2005
3. ICH Q10(R2) Pharmaceutical Quality System, ICH Harmonised Tripartite Guideline, International Conference on Harmonisation of Technical Requirements for Registration of Pharmaceuticals for Human Use, 2009
4. ICH Q11(R2) Development and Manufacture of Drug Substances (Chemical Entities and Biotechnological/Biological Entities), ICH Harmonised Tripartite Guideline, International Conference on Harmonisation of Technical Requirements for Registration of Pharmaceuticals for Human Use, 2012
5. Togkalidou, R. D. Braatz, B. K. Johnson, O. Davidson, and A. Andrews. Experimental design and inferential modeling in pharmaceutical crystallization. AIChE J., 47:160-168, 2001
6. Togkalidou, H.-H. Tung, Y. Sun, A. Andrews, and R. D. Braatz. Solution concentration prediction for pharmaceutical crystallization processes using robust chemometrics and ATR FTIR spectroscopy. Org. Process Res. Dev., 6:317-322, 2002
7. Togkalidou, H.-H. Tung, Y. Sun, A. Andrews, and R. D. Braatz. Parameter estimation and optimization of a loosely-bound aggregating pharmaceutical crystallization using in-situ infrared and laser backscattering measurements. Ind. Eng. Chem. Res., 43:6168- 6181, 2004
8. S. Hong, K. A. Severson, M. Jiang, A. E. Lu, J. C. Love, and R. D. Braatz. Challenges and opportunities of biopharmaceutical manufacturing control. Computers & Chemical Engineering, 110:106-114, 2018
9. Lakerveld, P. L. Heider, K. D. Jensen, R. D. Braatz, K. F. Jensen, A. S. Myerson, and B. L. Trout. End-to-end continuous manufacturing: Integration of unit operations. In Continuous Manufacturing of Pharmaceuticals, edited by P. Kleinebudde, J. Khinnast, and J. Rantanen, Wiley, New York, Chapter 13, pages 447-483, 2017
10. Mo Jiang and Richard D. Braatz. Integrated control of continuous (bio)pharmaceutical manufacturing. Am. Pharm. Rev., 19(6):110-115, 2016
11. Lakerveld, B. Benyahia, P. L. Heider, H. Zhang, A. Wolfe, C. J. Testa, S. Ogden, D. R. Hersey, S. Mascia, J. M. B. Evans, R. D. Braatz, and P. I. Barton. The application of an automated control strategy for an integrated continuous pharmaceutical pilot plant. Organic Process Research & Development, 19(9):1088-1100, 2015

12. Jiang, Y. E. D. Li, H. H. Tung, and R. D. Braatz. Effect of jet velocity on crystal size distribution from antisolvent and cooling crystallizations in a dual impinging jet mixer. Chemical Engineering and Processing: Process Intensification, 97:242-247, 2015
13. Jiang, Y. E. D. Li, H. H. Tung, and R. D. Braatz. Effect of jet velocity on crystal size distribution from antisolvent and cooling crystallizations in a dual impinging jet mixer. Chemical Engineering and Processing: Process Intensification, 97:242-247, 2015
14. Amos E. Lu, Joel A. Paulson, Nicholas J. Mozdzierz, Alan Stockdale, Ashlee N. Ford Versypt, Kerry R. Love, J. Christopher Love, and Richard D. Braatz. Control systems technology in the advanced manufacturing of biologic drugs. Proceedings of the IEEE Conference on Control Applications, Sydney, Australia, 1505-1515, 2015.
15. T. Zhang, R. Lakerveld, P. L. Heider, M. Y. Tao, M. Su, C. J. Testa, A. N. D'Antonio, P. I. Barton, R. D Braatz, B. L. Trout, A. S. Myerson, K. F. Jensen, and J. M. B. Evans. Application of continuous crystallization in an integrated continuous pharmaceutical pilot plant. Crystal Growth & Design, 14(5):2148-2157, 2014
16. Mascia, P. L. Heider, H. Zhang, R. Lakerveld, B. Benyahia, P. I. Barton, R. D. Braatz, C. L. Cooney, J. M. B. Evans, T. F. Jamison, K. F. Jensen, A. S. Myerson, and B. L. Trout. End-to-end continuous manufacturing of pharmaceuticals: Integrated synthesis, purification, and final dosage formation. Angewandte Chemie Int. Ed., 52(47):12359-12363, 2013
17. Lakerveld, B. Benyahia, R. D. Braatz, and P. I. Barton. Model-based design of a plant-wide control strategy for a continuous pharmaceutical plant, AIChE Journal, 59:3671-3685, 2013
18. C. S. Kee, R. B. H. Tan, and R. D. Braatz. Semiautomated identification of the phase diagram for enantiotropic crystallizations using ATR-FTIR spectroscopy and laser backscattering. Ind. Eng. Chem. Res., 50:1488-1495, 2011
19. C. S. Kee, P. D. Arendt, L. M. Goh, R. B. H. Tan, and R. D. Braatz. Nucleation and growth kinetics estimation for L-phenylalanine hydrate and anhydrate crystallization. CrystEngComm, 13:1197-1209, 2011
20. C. S. Kee, P. D. Arendt, R. B. H. Tan, and R. D. Braatz. Selective crystallization of the metastable anhydrate form in the enantiotropic pseudo-dimorph system of L-phenylalanine using concentration feedback control. Crystal Growth & Design, 9:3052-3061, 2009
21. C. S. Kee, R. B. H. Tan, and R. D. Braatz. Selective crystallization of the metastable alpha-form of L-glutamic acid using concentration feedback control. Crystal Growth & Design, 9:3044-3051, 2009
22. X. Zhou, M. Fujiwara, X. Y. Woo, E. Rusli, H.-H. Tung, C. Starbuck, O. Davidson, Z. Ge, and R. D. Braatz. Direct design of pharmaceutical antisolvent crystallization through concentration control. Crystal Growth & Design, 6:892-898, 2006

555508 Combining Big Data Analytics and Computational Chemistry in Target-Based Drug Discovery

A new target-based drug screening method is proposed exploiting the synergy effect of big data analytics and computational chemistry. The method was illustrated in the development of a target-based drug using 45 sulphonamide derivatives (33 training compounds and 12 testing compounds) with carbonic anhydrase IX (CA IX) as a drug target. For each sulphonamide compound, about 5,000 molecular descriptors were calculated using Gaussian (Gaussian Inc.), and lipophilicity ($logk_w$) and inhibitory activity ($logK_i$) were measured using RP-LC as drug properties. Genetic algorithm-partial least squares (GA-PLS) provided a quantitative structure-property-relationship (QSPR) model with high prediction capability employing only 7 descriptors. The ideal drug structure was obtained by inverting the QSPR model numerically in respect to reference properties of a reference drug (one of testing compounds). 12 testing compounds were ranked using the ideal drug structure and the rank was further validated through molecular docking and molecular dynamic (MD) simulation of drug candidate-CA IX complexes. Parameterization of the simulation was verified experimentally using structural analyses of the complexes including MALDI-TOF/TOF-MS. The results proved that the drug screening method proposed has great flexibility and can provide drug candidates with lower risk.

567740 Utilizing Data Systems and Modeling to Evaluate Equipment Fit and Technical Capabilities in External Drug Substance Manufacturing

Both clinical and commercial small molecule pharmaceutical intermediates and API are increasingly manufactured at Contract Manufacturing Organizations (CMO's) relative to internal manufacturing plants. This paradigm shift implies some loss of ability and flexibility to ensure optimal process fit within the plant equipment. Internal manufacturing plants allow for self-selection of equipment for process fit. At an external plant, the CMO selects the equipment for a process. Particularly in the clinical space where processes are new and CMO experience with the processes is low, this leads to a higher risk of poor equipment selection, potentially causing timeline delays, quality insufficiencies, and yield fluctuation. To mitigate this risk, our team developed data systems and a suite of tools to assess process fit and technical capabilities throughout our CMO network. For the data systems, we used a combination of vendor survey and historical operating experience to assemble a consolidated CMO equipment database and a plant capability log. Our suite of tools, including various Dynochem utilities, interactive data visualization, and statistical modeling, leverages our data systems and allows us to cross-reference CMO equipment and capabilities against individual API or intermediate processes. Typical products of this evaluation include proactive equipment fit, decision analysis around vendor selection, and evaluations of equipment impact on process performance. As a result, we can proactively address issues with vendor capabilities and equipment fit, which saves us from delays and cost increases. In this talk, we will describe the systems that we implemented, the workflows that were developed, and several case studies describing how we used these tools to address individual problems associated with technology transfer and scale up in an external manufacturing network for clinical API.

575528 Robust Artificial Neural Networks for Nonlinear Model Predictive Control of Multiscale Stochastic Systems

Multiscale stochastic systems are comprised of non-closed-form expressions that represent the discrete scale and continuum transport equations that describe macroscale phenomena. The discrete and continuum scales are coupled in order to provide the results of accurate microscale simulations to the domain sizes available to the macroscale simulations. However, the presence of non-closed-form expressions introduces noise into the observables and increases the total computational cost such that these multiscale models become too intensive for online model predictive control applications. This motivates the development of computationally efficient data-driven models to predict the responses of the full multiscale model to changes in the input parameters. Artificial Neural Networks (ANNs) are a machine learning technique that demonstrated the ability to accurately and efficiently predict the relationships between diverse input and output timeseries [1,2]. However, the ANN performance deteriorates when trained on noisy data [1] because the weights of the networks aim to capture the mean responses to the inputs. To reduce model-plant mismatch in the case of multiscale stochastic systems, it is necessary to predict the distribution of the response variables obtained from the ANNs rather than only their expectation values.

In our previous work [3], we used ANN models to predict the responses of a stochastic multiscale model of thin film formation by Chemical Vapour Deposition (CVD) [4]. The ANN models were subsequently incorporated into a shrinking horizon nonlinear model predictive control scheme. The predicted responses were roughness and growth rate while the manipulated variables were substrate temperature and bulk precursor mole fraction. It was demonstrated that the ANNs can efficiently provide accurate predictions of the system's observables and reject disturbances not seen during training. However, uncertainty was not considered when generating the data from the stochastic multiscale model.

In this work, we propose training robust ANNs with distributed weights under parametric uncertainty to efficiently generate predictions that account for the variability in the observables. As a case study, we employ the aforementioned CVD system [4] to generate the training datasets under parametric uncertainty. We obtain the distributions and the statistical moments of the observables and subsequently incorporate the robust ANNs into a nonlinear model predictive control framework. The predicted trajectories are subsequently validated against the multiscale CVD model. Rather than obtaining only the mean response of the observables to the manipulated variables, we predict several possible time series of the observables and provide bounds, thereby enabling robust control of the multiscale stochastic system in the presence of parametric uncertainty.

References

[1] Svozil, D., Kvasnicka, V. and Pospichal, J. Introduction to Multi-Layer Feed-Forward Neural Networks, Chemometrics and Intelligent Laboratory Systems, vol. 39, no. 1, pp. 43–62, Nov. 1997.

[2] Venkatasubramanian, V. The promise of artificial intelligence in chemical engineering: Is it here, finally?, AIChE Journal, vol. 65, no. 2, pp. 466–478, 2019.

[3] Kimaev, G. and Ricardez-Sandoval, L. A. Nonlinear Model Predictive Control of a Multiscale Thin Film Deposition Process Using Artificial Neural Networks, Chemical Engineering Science, submitted on 2019-Apr-06.

[4] Vlachos, D. Multiscale Integration Hybrid Algorithms for Homogeneous-Heterogeneous Reactors, AIChE Journal, vol. 43, no. 11, 1997.

577263 Self-Optimizing Control of a Continuous Pharmaceutical Manufacturing Plant

The pharmaceutical industry is increasing its use of continuous manufacturing, which enables easier scale-up, faster time-to-market, and tighter control over product quality [Plumb, 2005]. Continuous operation is well suited for the application of plantwide control approaches such as self-optimizing control (SOC), where the goal is to find a set of controlled variables (CVs) which, when kept at constant setpoints, indirectly lead to near-optimal operation [Skogestad, 2000, Jaschke et al., 2017].

The main idea of SOC is to control optimal invariants, which are process variables whose optimal setpoints are insensitive to disturbances and measurement noise. The ideal SOC variable would be the gradient of the objective function, where the necessary condition for optimality is enforced by controlling the gradient to zero. In real processes, where the gradient cannot be measured directly, selecting self-optimizing controlled variables is equivalent to finding good approximations of the gradient using single measurements or combinations, often linear, of measurements.

An SOC structure is able to track the optimal steady-state operating point of the plant closely in the presence of uncertainty, without the need for re-optimization. In hierarchical control structures which have time-scale separation between the supervisory control and optimization layers, SOC at the supervisory layer can ensure optimal operation when disturbances impact the process, without waiting for setpoint updates given by the upper level optimization layer. Only in the presence of unmodeled disturbances or when large disturbances move the process far away from the nominal operating point, a new setpoint is computed from the optimizer to correct for the steady-state loss of optimality.

This work explores how self-optimizing control can be applied to the continuous-flow synthesis of atropine, and appears to be the first application of self-optimizing control (SOC) to a continuous pharmaceutical manufacturing plant reported in the literature.

This study employs an improved version of a recently published first-principles model of the continuous-flow synthesis of atropine [Nikolakopoulou et al., 2019], which is an active pharmaceutical ingredient (API) with a variety of therapeutic uses, including the treatment of heart rhythm problems. The model was constructed from the process flowsheet and experimental results reported by [Bedard et al., 2017]. The Environmental factor (E-factor, defined by the ratio of the mass of waste per mass of product) is used as the objective function to find the optimal operating point of the plant. The main sources of uncertainty considered when designing the control structure include perturbations in the process variables (disturbances), parametric model uncertainty, and sensor noise.

The controlled variables are selected in two steps. First, local methods [Halvorsen et al., 2003, Kariwala et al., 2008, Alstad et al., 2009, Kariwala and Cao, 2009] are used to screen promising candidate controlled variables around the nominal operating point, where the plant is expected to operate most of the time. These local methods rely on a linearized model of the plant and a quadratic approximation of the objective function. The performance of a given control structure is quantified in terms of the loss of optimality, i.e., the difference between the value of the objective function resulting from constant setpoint control using that particular control structure and the value of the objective function resulting from truly optimal operation [Halvorsen et al., 2003]. The sets of controlled variables are selected by systematically minimizing the average loss of optimality with respect to the given objective function (in this case, the E-factor). In a second step, these preliminary candidate controlled variables are then validated over the entire operating envelope using the original nonlinear model of the plant and Monte Carlo simulations for independent and normally distributed scenarios of disturbances and measurement noise realizations.

It was found that near-optimal operation can be achieved by controlling a linear combination of flow rate and concentration measurements to constant setpoints, with only small losses of optimality for all of the model parametric uncertainties, disturbances, and measurement noise considered in this case study.

References

Alstad, V., Skogestad, S., and Hori, E. S. (2009). Optimal measurement combinations as controlled variables. *Journal of Process Control*, 19(1):138–148.

Bedard, A.-C., Longstreet, A. R., Britton, J., Wang, Y., Moriguchi, H., Hicklin, R. W., Green, W. H., and Jamison, T. F. (2017). Minimizing E-factor in the continuous-flow synthesis of diazepam and atropine. *Bioorganic & Medicinal Chemistry*, 25(23):6233–6241.

Halvorsen, I. J., Skogestad, S., Morud, J. C., and Alstad, V. (2003). Optimal selection of controlled variables. *Industrial & Engineering Chemistry Research*, 42(14):3273–3284.

Jaschke, J., Cao, Y., and Kariwala, V. (2017). Self-optimizing control – A survey. *Annual Reviews in Control*, 43:199–223.

Kariwala, V. and Cao, Y. (2009). Bidirectional branch and bound for controlled variable selection. Part II: Exact local method for self-optimizing control. *Computers & Chemical Engineering*, 33(8):1402–1412.

Kariwala, V., Cao, Y., and Janardhanan, S. (2008). Local self-optimizing control with average loss minimization. *Industrial & Engineering Chemistry Research*, 47(4):1150–1158.

Nikolakopoulou, A., von Andrian, M., and Braatz, R. D. (2019). Supervisory control of a compact modular reconfigurable system for continuous-flow pharmaceutical manufacturing. In *Proceedings of the American Control Conference*. in press.

Plumb, K. (2005). Continuous processing in the pharmaceutical industry: Changing the mind set. *Chemical Engineering Research Design*, 83(6):730–738.

Skogestad, S. (2000). Plantwide control: The search for the self-optimizing control structure. *Journal of Process Control*, 10(5):487–507.

569159 Integration of Design, Scheduling, and Control of Batch Processes By Model Based Multiparametric Programming

The complexity of decision making problems in the process industry has conventionally resulted in isolation of decisions with respect to the time scales of their impact on the operation. These long term (design), middle term (scheduling), and short term (control) decisions are typically structured hierarchically with an information flow allowed dominantly in descending order in their time scales [1]. Integration of these layers across an enterprise is expected to improve the process economics and reliability by benefiting from the synergistic interactions between different decisions [2]. However, the literature hitherto has focused on continuous processes, whereas the batch processes impose further challenges for the complete integration due to the high degrees of freedom in the scheduling problem introduced by the multipurpose/multiproduct units [3-5].

In this work, we present a unified theory and framework to integrate the process design, scheduling, and control decisions based on a single high fidelity model. We develop explicit strategies for (i) multiparametric rolling horizon optimization (mpRHO) for middle term economical decisions as a function of closed-loop states and time-variant market conditions, and (ii) multiparametric model predictive control (mpMPC) to effectively track the set-points determined by the mpRHO . The offline nature of these operational strategies allows for their direct implementation in (i) the dynamic high-fidelity model of the batch process, as well as (ii) a mixed-integer dynamic optimization formulation for the optimal design configuration simultaneously with the scheduling and control problems [1]. The introduced framework will be demonstrated on a flowshop batch process with multiple end-products.

1. Burnak, B., Diangelakis, N.A., Katz, J., Pistikopoulos, E.N., Integrated process design, scheduling, and control using multiparametric programming, Computers & Chemical Engineering (2019), 125, pp. 164-184.
2. Terrazas-Moreno, S., Flores-Tlacuahuac, A., Grossmann, I.E., Simultaneous design, scheduling, and optimal control of a methyl-methacrylate continuous polymerization reactor, AIChE J. (2008), 54 (12), pp. 3160-3170.
3. Patil, B.P., Maia, E., Ricardez-Sandoval, L.A., Integration of scheduling, design, and control of multiproduct chemical processes under uncertainty, AIChE J. (2015), 61 (8), pp. 2456-2470.
4. Zhuge, J., Ierapetritou, M.G., Integration of scheduling and control for batch processes using multi-parametric model predictive control, AIChE J. (2014), 60 (9), pp. 3169-3183.
5. Rossi, F., Casas-Orozco, D., Reklaitis, G., Manenti, F., Buzzi-Ferraris, G., A computational framework for integrating campaign scheduling, dynamic optimization and optimal control in multi-unit batch processes, Computers and Chemical Engineering (2017), 107, pp. 184-220.

570308 A Systematic Model-Based Approach for the Design and Control of Protein Crystallization

Because of its high resolution, packed-bed chromatography is the most widely used method for separation in bioprocesses. However, this purification method has an operating cost, roughly the cost of resin divided by the number of times that the resin can be reused, that scales linearly with throughput. As bioreactors have become more productive, the operating costs of chromatography have become a significant portion of the total costs of manufacturing. On the other hand, crystallization from liquid solution has cost that scales sub-linearly with throughput, which is why this method is heavily used for the purification of amino acids, active pharmaceutical ingredients, and intermediates. Although crystallization is used for the production-scale purification of some therapeutic proteins such as insulin, research and development are needed to develop technology effective for most therapeutic proteins [1].

Technology for the design and control of the crystallization of proteins is much less mature than for small molecules. Methods for inducing supersaturation are also more limited due to need of maintaining protein stability and quality. Many proteins are easily denatured by changes in temperature and pH, addition of precipitants, and agitation. Proteins have complex thermodynamics, slow kinetics, large uncertainties, and potential for protein aggregation that greatly restrict allowable paths through the phase diagram, which is equivalent to threading an unknown narrow winding path through an uncertain high-dimensional space.

Although numerous studies on protein crystallization have been published, most of them performed scale-up from micro-batch scale experiments to medium-scale batch crystallizers without modeling the crystallization kinetics. The presentation describes a systematic approach to the design and control of protein crystallization from droplet-scale to medium-scale by a combination of first-principles models and model-based optimization.

Preliminary well- or vial-based experimental data can provide lower bounds on nucleation rates. Most experimental studies apply screening methods like hanging-drop or sitting-drop vapor diffusion systems [2]. Nucleation within such drops is describable by a stochastic model in the form of a Master equation [3]:

$$dP_0(t)/dt = -B_0(t)V(t)P_0(t),\ P_0(0)=1,$$

where B_0 is the nucleation rate which is a function of states that change with time (more details below), P_0 is the time evolution of the probability that a droplet contains no crystals, and V is the volume of the droplet. A lower bound for the nucleation rate can be obtained by comparing mean induction time calculated from the model and experimental results.

The quantity of the protein for carrying out preliminary experiments for the estimation of crystallization kinetics and thermodynamics is very limited during the initial stages of process development. This observation motivated our development of a droplet-based evaporative crystallizer for the evaluation of candidate crystallization conditions obtained from publications and/or preliminary studies, and for the estimation of crystallization kinetics with minimum quantity of protein. Controlled evaporation and rehydration of the droplet are enabled by feeding air of controlled temperature and humidity into the chamber containing the droplet. Real-time image data are collected using a microscope through an optically visible chamber. This system is describable by a crystal population balance and solution mass balances:

$$\partial f(L,t)/\partial t = J(t)f(L,t)/V(t) + B_0(S,T,\mu_2)\delta(L) - G(S,T)\partial f(L,t)/\partial L,$$

$$dC_P(t)/dt = J(t)C_P(t)/V(t) - 3\rho_c k_v G(S,T)\mu_2(t),$$

$$dC_B(t)/dt = J(t)C_B(t)/V(t),$$

$dV(t)/dt = -J(t)$,

where f is the crystal size distribution, L is the characteristic length, G is the crystal growth rate, B_0 is the crystal nucleation rate (which can be a combination of primary and secondary nucleation), J is the evaporation rate, C_P is the protein concentration in solution, C_B is the concentration of buffer/precipitants, S is the supersaturation (which is a function of the solubility and the solution concentrations), T is the temperature, and μ_2 is the second-order moment of the crystal size distribution. Real-time data during evaporation and rehydration allow the estimation of crystallization kinetics and yield. Furthermore, controlled evaporation and rehydration allow exploration of the phase diagram and evaluation of the potential path in the scaled-up crystallization.

For scale-up, a mixed-tank design of a medium-scale continuous evaporative crystallizer was developed. The pressure in the crystallizer is lowed by a vacuum pump to induce evaporation, and real-time imaging and product collection are enabled through the tube leaving the crystallizer controlled by a valve. A model for this system, which is similar to the droplet-based system, is

$$\partial f(L,t)/\partial t = (J(t)-F_{in}(t))f(L,t)/V(t) + B_0(S,T,\mu_2)\delta(L) - G(S,T)\partial f(L,t)/\partial L,$$

$$dC_P(t)/dt = F_{in}(t)(C_{P,in}(t)-C_P(t))/V(t) + J(t)C_P(t)/V(t) - 3\rho_c k_v G(S,T)\mu_2(t),$$

$$dC_B(t)/dt = F_{in}(t)(C_{B,in}(t)-C_B(t))/V(t) + J(t)C_B(t)/V(t),$$

$$dV(t)/dt = F_{in}(t) - F_{out}(t) - J(t),$$

where F_{in} and F_{out} are the volumetric flow rates of protein solution entering and leaving the crystallizer. Estimated model parameters from the droplet-based crystallization system are used to design optimal controlled operations for the medium-scale crystallizer.

The proposed overall systematic approach for the design and control of protein crystallization based on very small quantities of protein is evaluated in a detailed simulation case study using phase equilibria and crystallization kinetics experimentally reported in the literature for lysozyme [4]–[7], which is an antibacterial protein.

References:

[1] M. S. Hong, K. A. Severson, M. Jiang, A. E. Lu, J. C. Love, and R. D. Braatz, Challenges and opportunities in biopharmaceutical manufacturing control, Comput. Chem. Eng., vol. 110, pp. 106-114, 2018.

[2] A. McPherson, Introduction to protein crystallization, Methods, vol. 34, no. 4, pp. 254-265.

[3] L. Goh, K. Chen, V. Bhamidi, G. He, N. C. Kee, P. J. Kenis, C. F. Zukoski III, and R. D. Braatz, A stochastic model for nucleation kinetics determination in droplet-based microfluidic systems, Cryst. Growth Des., vol. 10, no. 6, pp. 2515-2521, 2010.

[4] D. Rosenbaum, P. C. Zamora, and C. F. Zukoski, Phase Behavior of Small Attractive Colloidal Particles, Phys. Rev. Lett., vol. 76, no. 1, pp. 150-153, 1996.

[5] M. V. Saikumar, C. E. Glatz, and M. A. Larson, Lysozyme crystal growth and nucleation kinetics, J. Cryst. Growth, vol. 187, pp. 277-288, 1998.

[6] E. L. Forsythe, R. A. Judge, and M. L. Pusey, Tetragonal chicken egg white lysozyme solubility in sodium chloride solutions, J. Chem. Eng. Data, vol. 44, no. 3, pp. 637-640, 1999.

[7] D. Shi, P. Mhaskar, N. H. El-Farra, and P. D. Christofides, Predictive control of crystal size distribution in protein crystallization, Nanotechnology, vol. 16, no. 7, pp. S562-S574, 2005.

574577 Multi-Model Based Model Predictive Control of a Rotational Molding Process

Rotational Molding, also known as rotomolding or rotational casting, is a batch process used for the processing of plastics through distinct heating and cooling phases [1]. In this process a mold filled with a powdered charge is slowly rotated in a heated oven, resulting in the softened material being dispersed and sticking to the walls of the mold. In order to maintain an even wall thickness, the mold is continuously rotated at all times both during the heating and cooling phase. One of the control challenges is to obtain consistent high-quality products from one batch to another while avoiding unfinished parts with incomplete curing or degradation due to prolonged overheating.

In a recent work [2], a batch subspace identification [3] based state-space modeling and control approach is used for rotational molding to achieve desired quality specifications. The modeling approach facilitates the accommodation of variable duration batches without the need for alignment of batch lengths as demonstrated in several other applications as well [3-6]. The results demonstrate the merits of the approach in terms of improved quality, variability rejection across difference batches and the ability to achieve the specified grade of product. In that work, however, the data-driven model is built only for the heating phase. Furthermore, while the implementation tests its merit to achieve new quality targets, the ability of the modeling and control framework to reject raw material variability is not explicitly demonstrated.

Motivated by these considerations, this work presents a multi-model-based model predictive control of a uni-axial rotational molding process to achieve the desired quality specifications. The model consists of two deterministic models for the heating phase and two stochastic models for the cooling phase, identified using deterministic and stochastic batch subspace identification respectively. The four models corresponding to these sequential stages in the process are connected together using the least squares model between the terminal states of the previous model and initial states of the next model. This modeling approach results in improved accuracy of the model predictions enabling improved control of the product properties. In particular, in the previous work [2], the cooling cycle was terminated upon reaching a predefined internal mold temperature. In contrast, in the present work the cooling cycle is terminated based on the real-time predictions such that the final product of desired specifications is achieved. Further, to evaluate the robustness of the proposed approach, we blend the original polymer resin with a similar resin with slightly different melt index and demonstrate the meeting of product specification.

[1] Gomes, F. P. C., Garg, A., Mhaskar, P. and Thompson, M. R. Thompson, Quality monitoring of rotational molded parts using a nondestructive technique, ANTEC (2018).

[2] Garg, A., Gomes, F. P. C., Mhaskar, P., & Thompson, M. R. (2019). Model predictive control of uni-axial rotational molding process. Computers & Chemical Engineering, 121, 306-316.

[3] Corbett, B., Mhaskar, P., Subspace identification for data-driven modelling and quality control of batch processes, AIChE Journal 62 (2016) 1581-1601.

[4] Garg, A., Mhaskar, P. (2017), Subspace Identification Based Modeling and Control of Batch Particulate Processes, Industrial and Engineering Chemistry Research, 56 (26) 7491-7502.

[5] Garg, A., Corbett, B., Mhaskar, P., Hu, G., Flores-Cerrillo, J. (2017), High Fidelity Model Development and Subspace Identification of a Hydrogen Plant Startup Dynamics, Computers and Chemical Engineering, 106 183-190.

[6] Rashid, M. M., Mhaskar, P., & Swartz, C. L. (2017). Handling multi-rate and missing data in variable duration economic model predictive control of batch processes, AIChE Journal, 63(7), 2705-2718.

575174 Multiscale Modeling and Control of Cell Wall Thickness in Batch Pulp Digester

In pulping processes, the cell wall thickness (CWT) of wood chips is gradually decreased due to the removal of lignin, which causes the cellulose fiber to collapse. Paper made of collapsed and uncollapsed fibers has very different paper properties (e.g., density, surface smoothness and ink holdout capability). For example, the collapsed fiber is suitable for high-quality printing paper as it has dense and smooth surface whereas the uncollapsed fiber is better for making cartonboard that is resistant to bending and tearing [1-2]. However, the evolution of this important microscopic paper property is not able to be described by existing mathematical models as they only focus on macroscopic properties such as the concentration profiles of components and the temperature profiles of the cooking liquor and wood chip [3-6].

Motivated by this limitation, we developed a novel multiscale model by combining the mass continuity and thermal energy balance equations adopted from a modified "extended Purdue model" [6] with a kinetic Monte Carlo algorithm [7-9] to describe the microscopic events such as the evolution of CWT and Kappa number (i.e., residual lignin content in the wood pulp). To handle the high computational cost of the proposed multiscale mode, a data-driven reduced-order model was developed by utilizing the multivariable output error state space (MOESP) algorithm and the high-fidelity input/output data [10]. In addition, as the primary measurements of the process (i.e., Kappa number and CWT) are not available in real-time, a soft sensing system (i.e., Kalman filter) was developed to estimate the primary measurements by utilizing the secondary measurements (e.g., active effective alkali and dissolved lignin concentrations in the free-liquor phase) [11]. By taking advantages of the soft sensing system, a model-based feedback control framework was developed to regulate both the CWT and Kappa number of wood chips. Specifically, a model predictive control framework is employed to find the optimal manipulated input sequence (i.e., the free-liquor temperature of pulp digester) that drives the controlled output variables (i.e., primary measurements) to the pre-specified target values. The closed-loop simulation results demonstrated the proposed control framework outperforms other existing control strategies in regulating the microscopic paper properties.

References

[1] Johansson A. Correlations between fibre properties and paper properties. *Master's thesis,* KTH Royal Institute of Technology, Stockholm, Sweden, **2011.**

[2] Facada MJ. Influence of Kraft paper quality on the performance of an industrial paper impregnation process. *Ph.D. thesis*, Universidade Tecnica de Lisboa, Lisboa, Portugal, **2015.**

[3] Funkquist J. Grey-box identification of a continuous digester – a distributed-parameter process. *Control Eng. Pract.*, **1997**, *5*, 919-930.

[4] Wisnewski PA.; Doyle FJ. Fundamental continuous pulp-digester model for simulation and control. *AIChE J.*, **1997**, *43*, 3175-3192.

[5] Smith C. Studies of the mathematical modeling simulation and control of the operation of Kamyr continuous digester for the Kraft process. *Ph.D. thesis,* Purdue University, West Lafayette, IN, **1974**.

[6] Bhartiya S.; Dufour P.; Doyle FJ. Fundamental thermal-hydraulic pulp digester model with grade transition. *AIChE J.*, **2003**, *49*, 411-425.

[7] Lee D.; Mohr A.; Kwon JSI. Kinetic Monte Carlo modeling of multivariant binding of CTB proteins with GM1 receptor. *Comp. Chem. Eng.,* **2018**, *118*, 283-295.

[8] Kwon JSI.; Nayhouse M.; Christofides PD.; Orkoulas G. Protein crystal shape and size control in batch crystallization: Comparing model predictive control with conventional operating policies. *Ind. Eng. Chem. Res.*, **2014**, *53*, 5002-5014.

[9] Choi HK.; Kwon JSI. Multiscale modeling and control of Kappa umber and porosity in a batch-type pulp digester. *AIChE J.* (In press), **2019**.

[10] Chou CT.; Verhaegen M. Subspace algorithms for the identification of multivariable dynamic errors-in-variables model. *Automatica,* **1997**, *33*, 1857-1869.

[11] Galicia HJ.; He QP.; Wang JA. A reduced order soft sensor approach and its application to a continuous digester. *J. Process Control.*, **2011**, *21*, 489-500.

566999 Modeling and Uncertainty Analysis of Microbial Electrosynthesis of Acetate for Martian Colonization

The primary goal of the Center for Utilization of Biological Engineering in Space (CUBES) is efficient exploitation of Martian *in-situ* resources for integrated bio-manufacturing of the necessary products for Martian colonization, including biologically-produced pharmaceuticals, cell-based treatments/therapeutics, and materials for on-demand diverse additive manufacturing applications [1]. A key module of the bio-manufacturing system is a novel, biologically-driven process for carbon fixation through microbial electrosynthesis, in which acetate is generated as a product of microbial growth. Acetate is an essential carbon source for downstream bioplastics and fuel production.

In this work, we present a framework for modeling a continuous bioreactor for carbon fixation through microbial electrosynthesis. An important requirement of the bioreactor model is the ability to reliably forecast the acetate synthesis under climate conditions of Mars and operational uncertainties within the integrated biomanufacturing system. We first present a first-principles dynamic model for a reverse microbial fuel cell reactor, where the bacteria on the cathode surface reduce CO_2 into acetate via a supply of electrons [2]. The underlying physical phenomena include coupled micro-scale reaction and transport processes within the biofilm, along with macro-scale mass balances in the bulk liquid [3]. The reaction kinetics are described by the Butler Volmer model for the anode and the Butler-Volmer-Monod model for the cathode [4]. The cathode surface potential is treated as a time-dependent variable related to the efficiency of a photovoltaic panel absorbing light incident to the Martian surface. The climate conditions on Mars due to dust and cloud formations affect the availability of light, acting as a key disturbance to the production rate of acetate. We utilize the RedSun climate modeling software, developed within CUBES [5], to provide a description for the state of the Martian climate and *in-situ* variables such as temperature, pressure, and the spectral flux based on a geospatial grid of Mars.

For a nominal period of a human mission on Mars, we present a comprehensive parametric study on the system behavior to assess the effects of various reactor design and mission parameters on the quantities of interest, such as the biofilm growth rate and acetate production. The aforementioned findings inform the practical engineering and operational decisions of the integrated biomanufacturing system. Finally, we utilize deep learning methods [6] to develop a *surrogate model* for the first-principles bioreactor model in order to systematically study the performance of the bioreactor under the varying climate conditions of Mars as well as operational uncertainties of the biomanufacturing system.

References

[1] Menezes A., A., Cumbers J., Hogan J.A., Arkin A., "Towards synthetic biological approaches to resource utilization on space missions", Journal of the Royal Society Interface, 2015, 12 (102), 20140715.

[2] Kazemi, M., Biria , D., Rismani-Yazdi, H., "Modelling bio-electrosynthesis in a reverse microbial fuel cell to produce acetate from CO_2 and H_2O", *Phys.Chem.Chem.Phys.*, 2015, 17, 12561

[3] International Water Association Task Group on Biofilm Modeling, "Mathematical Modeling of Biofilms", IWA Publishing, Water Intelligence Online, 2006

[4] Hamelers H.V.M, Ter Heijne A., Stein N., Rozendal R.A., Buisman C., J., N., "Butler–Volmer–Monod model for describing bio-anode polarization curves", *Bioresource Technology*, 2011, 102, 381-387

[5] "Abel A., Berliner A.J., Mirkovic M., Collins W., Arkin A.P., Clark D.S, "Traversing Photovoltaic and Photoelectrochemical Production Capacity Across the Martian Surface", to be submitted for publication.

[6]Tripathy R. K. , Bilionis I., Wiart J., "Deep UQ: Learning deep neural network surrogate models for high dimensional uncertainty quantification",*Journal of Computational Physics*, 2015, 375, 565-588.

579939 Data Science and Machine Learning in the Upstream Oil and Gas Industry

Digital technologies have radically transformed many industries in the last two decades, but advances in pervasive data access, machine learning, and digital business models have penetrated the upstream oil and gas industry more slowly. This process is now intensifying, with digital technologies poised to disrupt core business and technical workflows, in production operations, drilling, and subsurface characterization, among other areas. I will review some specific instances where digital technologies, especially those enabled by data science and machine learning, have been successfully utilized within ExxonMobil, drawing lessons for the effective development and deployment of these technologies to achieve transformational business impact.

579937 Big Data Analytics in the Advanced Manufacturing of Biopharmaceuticals

This presentation describes the use of process data analytics and machine learning to improve process productivity, reliability, and control in next-generation biopharmaceuticals manufacturing systems. New information streams are described, including the use of hyperspectral imaging to generate gigabytes of real-time data in the form of high order tensors, and the coupling of on-line automated sampling systems with liquid chromatography-mass spectroscopy (LC-MS) to simultaneously measure transient variations in large numbers of distinct small molecules and proteins during manufacturing. Experimental data are presented from multiple laboratories for a variety of bioprocess unit operations, including bioreactors and freeze-drying systems. The use of that data in first-principles, data-driven, and hybrid models are described, with relationships drawn to chemometrics (e.g., partial least squares, principal component analysis), system identification (e.g., canonical variate analysis), pattern recognition (e.g., Fisher and quadratic discriminant analysis), and machine learning (e.g., ensemble learning, elastic net regularization).

581227 Big Data Impact on the Future of Manufacturing

There is an opportunity for innovation in chemical and pharmaceutical manufacturing processes through the integration of human intuition and experience with big data, machine learning and other software innovations. Subject matter experts bring their experience and expertise for constructing an effective analytics strategy, whereas new capabilities in software enable them to work and analyze processes and outcomes at the speed of thought. This speed of insight is increasingly important in dealing with two of the highest priority pressure points in the industry: time to market from research to production, and replacing batch-based production with a continuous production model.

However, since analysis can suffer from human biases and heuristics, it is imperative to take advantage of big data platforms and applications to ensure completeness of the analytics effort. This includes data collection and aggregation for comprehensive access to production results, as well as advanced analytics to accelerate the time to insight and the breath of questions which may be required for insight. This session will provide a framework for leveraging software innovation in the context of industry requirements with examples of current customer deployments and insights.

580269 Data-Assisted Modeling and Optimization for Process Systems Engineering

Coupling equation-based first-principle modeling with information that comes in the form of input-output data for enhanced decision-making, is gaining significant interest in process systems engineering [1]. Most data-driven optimization techniques consider the system under study entirely as a "black-box", however, in order to take advantage of years of obtained physical process knowledge, this work proposes a hybrid modeling and optimization approach [2-3]. This talk will demonstrate the performance of a suite of flexible modeling and optimization tools that are able to incorporate both physics-based and machine-learning components, to enable reliable data-assisted decision making. Specifically, adaptive and tractable surrogate functions will be developed to represent input-output data, and these will be merged with mechanistic equations in the form of constraints or discrepancy modeling [4]. Optimization is performed via a custom-based branch-and-bound framework that incorporates sampling, machine learning and bounding of the uncertainty of the approximations. Finally, the performance of the proposed methods is demonstrated through a set of process design case studies that rely on multifidelity models and data.

References:

1. Venkatasubramanian, V. (2019), AIChE J., 65: 466-478
2. Boukouvala, F., Hasan, MMF, Floudas, C.A. (2017). JOGO, 67(1-2):3-42
3. Kim, S.H. and Boukouvala, F. (2019), Optimization Letters, 10.1007/s11590-019-01428-7
4. Kennedy, M. C. and O'Hagan, A. (2001), Journal of the Royal Statistical Society, 63: 425-464

576239 Pharmaceutical Process Intensification Via Continuous Manufacturing and the Role of Modeling and Simulation

The continuous manufacturing in the pharmaceutical industry for drug substance synthesis and drug product processing are shifting the manufacturing paradigm. The significant impact of process intensification and continuous manufacturing have been demonstrated for reducing energy, cost, time, waste, and footprint, but improving quality and consistency in quality, and also enabling new type of reactions (by flow-chemistry) and new paradigm in process control. Predictive models can aid process development by estimating the impact of process and equipment parameters, and material attributes (i.e., the model inputs) on product attributes (i.e., the model outputs) thereby providing a quantitative framework for assessing risk and evaluating risk mitigation approaches. In continuous manufacturing multiple synthetic steps can be operated in a single uninterrupted network of reactors. In continuous manufacturing, the dynamic interactions between different unit operations increases the complexity of the system.

Mechanistic models can describe process non-linearities and can be used to control the process and mitigate quality risks. In this work a comprehensive case study is presented, in which first principle models are developed for quality risk assessment of manufacturing processes. Examples of simulation application for drug substance and drug product processes will be discussed. And, the role of modeling and simulation in pharmaceutical process intensification via continuous manufacturing will be demonstrated by steady state and dynamic models.

Disclaimer: This scientific publication reflects the views of the authors and should not be construed to represent FDA's views or policies.

576026 Modular Process Intensification for Carbon Capture Using Data Driven Optimization

Modular chemical process intensification (PI) has attracted significant interest in the recent literature. Unlike conventional large-scale facilities, modular PI involves combining unit operations that usually appear together or dividing a unit into smaller subunits to enhance a specific function [1]. A modular facility consists of modules that can be easily added or removed, allowing more flexibility in capacity, product type, and geographic location [2]. Carbon capture is one area that can benefit from modularization. Depending on the source of emission, there exists a variability in the flue gas composition. For instance, post-combustion flue gas from coal-fired power plants has CO_2 concentrations ranging between 12.6-14%, while that of steel plants has a concentration ranging from 20-24% [3]. Furthermore, depending on the size and location of chemical plants, the modular design can be easily adjusted to handle different capacity and change in climate. Since most separation technologies only target a certain range of carbon content in flue gas under specific operating conditions, the performance and cost of carbon capture are largely affected by these variabilities. Thus, the identification of optimal modular design is crucial.

Optimization-based process synthesis can be used to determine the optimal modular design based on a given operating condition. Process synthesis involves developing a process superstructure and determining the optimal process configurations and parameters to achieve a certain process specification, such as minimizing energy requirement or maximizing profit. Process synthesis can be formulated into a mixed-integer nonlinear problem (MINLP) to allow selection of modules as well as optimal operating conditions. Accounting for all complex interactions between design decisions, optimization-based process synthesis is superior to heuristics-based process synthesis. However, one challenge of optimization-based process synthesis is that the resulting optimization problem is usually difficult to solve, especially when highly-complicated and accurate simulations are used to represent a module. The use of surrogate-based optimization has been proposed to reduce the complexity of the optimization problem [4, 5]. Surrogate-based optimization involves replacing complex computer simulations with surrogate models; the surrogate model is then optimized to determine the best modular design.

In this work, we propose an optimization-based modular PI strategy for carbon capture. We will specifically study modular vacuum pressure-swing adsorption (VPSA) with metal organic framework (MOFs). Several different types of MOFs will be integrated into modules, and the optimal modular design will be determined to address different variabilities intrinsic to carbon capture process, such as flue gas composition, geographic location of the plant, and desired capacity of the carbon capture facility. These variabilities will be analyzed using an actual CO_2 production data. Several factors of carbon capture process, such as purity, recovery, energy requirement, and productivity, will be used to analyze the performance of the modular design. The optimal modular design will be determined using surrogate-based optimization to reduce the complexity of the optimization problem.

References

1. Schug, B.W. and M.J. Realff, *Design of standardized, modular, chemical processes.* Computers & Chemical Engineering, 1996. **20**: p. S435-S441.
2. Baldea, M., et al., *Modular manufacturing processes: Status, challenges, and opportunities.* AIChE Journal, 2017. **63**(10): p. 4262-4272.
3. Hasan, M.M.F., et al., *Modeling, Simulation, and Optimization of Postcombustion CO2 Capture for Variable Feed Concentration and Flow Rate. 1. Chemical Absorption and Membrane Processes.* Industrial & Engineering Chemistry Research, 2012. **51**(48): p. 15642-15664.
4. Caballero, J.A. and I.E. Grossmann, *An algorithm for the use of surrogate models in modular flowsheet optimization.* AIChE Journal, 2008. **54**(10): p. 2633-2650.
5. Henao, C.A. and C.T. Maravelias, *Surrogate-based superstructure optimization framework.* AIChE Journal, 2011. **57**(5): p. 1216-1232.

574663 Process Simulation Study of Sour Gas Sweetening Technologies for Distributed Resources

Gas desulfurization is a very important step in utilization of natural gas resources, not only in preventing the formation of SO_2 after combustion, but also in avoiding poisoning of catalysts. However, for most distributed gas wells, the sour gas is directly burned and emitted into the atmosphere, which is neither environmental friendly nor economic. The estimated amount of methane vented or flared from the oil and gas industry as well as from landfills in the U.S. in 2012 was 213 billion ft^3 and 243 billion ft^3, respectively [1]. It is urgently needed to identify a suitable desulfurization process that can be used in distributed gas wells. Due to the varying conditions of these types of wells, the process should be flexible, low cost and operable at small scales. There are numbers of gas sweetening processes, such as liquid phase oxidation process, absorption and catalytic conversion [2]. Unfortunately, most of these processes are either designed for large scale (Claus process) or have very high energy cost (Amine system). Some processes, such as LOCAT [3] and scavenger [4] have the potential to be built in a modularized unit and operated in remote gas-containing resources. IntraMicron, Inc., an Auburn University. spin-off and small business located in Auburn, AL, has developed a low-cost desulfurization technology called SourCat to sweeten sour gas resources [1], which has lower desulfurization cost than the previous mentioned two processes.

In this talk, three desulfurization processes are selected (LOCAT, SourCat and scavenger) and simulated at different conditions. Then, an economic analysis tool (ECON) is used to calculate capital and operating cost. Comparing with real data ($250-1000 per ST range, LOCAT) [3, 4], we will show that our simulation results are representative of the real cases. Based on this, we can calculate the desulfurization cost for each process at different conditions. According to preliminary results, the SourCat process will offer approximately a 53% savings per MSCF compared to the LOCAT desulfurization cost at 100 MSCFD and 500 ppm H2S. Finally, a desulfurization cost plot is generated to provide guidance with respect to the preferred operating conditions of these three processes. Furthermore, based on different scales, a comparison of Aspen Economic Analyzer and ECON is performed to verify that ECON shows preference for small scale plants and Aspen Economic Analyzer prefers large scale systems. Finally, a process synthesis and intensification framework is considered by combining these processes, which has the potential to operate at larger scale with lower desulfurization cost.

[1] https://www.aiche.org/resources/publications/cep/2017/january/catalyzing-commercialization-sour-gas-has-sweeter-future

[2] Kohl A L, Nielsen R. Gas purification [M]. Elsevier, 1997.

[3] https://www.digitalrefining.com/article/1001502,Developments_in_H2S_scavenging.html#.XK-Eg5hKi72

[4] http://www.merichem.com/LO-CAT-Flexible-H2S-Removal-Process

[5] William Echt, "Liquid Redox Processes", Laurance Reid Gas Conditioning Conference: Fundamentals of Low-tonnage Sulfur Removal and Recovery, Norman, Oklahoma, February 2017.

[6] Bob Veroba, P. Eng., Erik Stewart, "Fundamentals of Gas Treating", Laurance Reid Gas Conditioning Conference, Norman, Oklahoma, February 2003.

569965 Process Intensification Framework for Reactive Separation Systems

Reactive separation (e.g., reactive distillation, membrane reactor, reactive absorption), which integrates multiple processing tasks into a single unit to maximize the synergy between multifunctional phenomena, is one of the major process intensification (PI) pathways towards breakthrough improvements [1-2]. As a classic example, significant energy and cost savings (15-80%) have been reported in reactive distillation processes over their conventional reactor-distillation counterparts [3].

Recent advances in the synthesis of reactive separation systems have been leveraging phenomena-based representation methods to come up with "out-of-the-box" PI process solutions beyond traditional unit operations [4-6]. However, the operational performances in these intensified designs under uncertainty and disturbances are mostly neglected at this synthesis stage, whereas their highly integrated schemes often decrease the degrees of freedom of the online decision maker, adversely affecting the process safety and limiting the operability of the systems [7]. Therefore, a holistic synthesis approach to derive such intensified systems with guaranteed safety, operability, and controllability performances is still lacking.

In this work, we propose a systematic process intensification framework to deliver operable reactive separation systems, which features: (i) phenomena-based process synthesis with the Generalized Modular Representation Framework (GMF) [8] to derive intensified reaction and/or separation design configurations, (ii) integrated flexibility and risk considerations to ensure steady-state operability and inherent safety performances [9], (iii) flowsheet identification and validation to translate the resulting phenomenological modular structure to intensified equipment/flowsheet, (iv) explicit model-based predictive controller design following the PAROC (PARametric Optimization and Control) framework [10] to ensure controllability under dynamic operation, and (v) simultaneous design and control with dynamic optimization strategies to close the loop for the design of verifiable, operable, and optimal reactive separation systems. The capabilities of the proposed framework are demonstrated with a case study on the production of methyl tertiary-butyl ether (MTBE), where design alternatives are systematically derived with guaranteed operability, safety, and controllability performances. Multiple process solutions can also be generated to show the tradeoffs between cost and operability performances for further decision making.

Reference

1. Bielenberg, J., & Palou-Rivera, I. (2019). The RAPID Manufacturing Institute – Reenergizing US Efforts in Process Intensification and Modular Chemical Processing. *Chemical Engineering and Processing-Process Intensification*.
2. Tian, Y., Demirel, S. E., Hasan, M. M. F., & Pistikopoulos, E. N (2018). An Overview of Process Systems Engineering Approaches for Process Intensification: State of the Art. *Chemical Engineering and Processing: Process Intensification*, 133, 160-210.
3. Harmsen, G. J. (2007). Reactive distillation: the front-runner of industrial process intensification: a full review of commercial applications, research, scale-up, design and operation. *Chemical Engineering and Processing: Process Intensification*, *46*(9), 774-780.
4. Demirel, S. E., Li, J., & Hasan, M. F. (2017). Systematic process intensification using building blocks. *Computers & Chemical Engineering, 105*, 2-38.
5. Tula, A. K., Babi, D. K., Bottlaender, J., Eden, M. R., & Gani, R. (2017). A computer-aided software-tool for sustainable process synthesis-intensification. *Computers & Chemical Engineering, 105*, 74-95.
6. da Cruz, F. E., & Manousiouthakis, V. I. (2016). Process intensification of reactive separator networks through the IDEAS conceptual framework. *Computers & Chemical Engineering, 105*, 39-55.
7. Baldea, M. (2015). From process integration to process intensification. *Computers & Chemical Engineering*, *81*, 104-114.
8. Papalexandri, K. P., & Pistikopoulos, E. N. (1996). Generalized modular representation framework for process synthesis. *AIChE Journal*, *42*(4), 1010-1032.

9. Tian, Y., & Pistikopoulos, E. N. (2019). Synthesis of Operable Process Intensification Systems—Steady-State Design with Safety and Operability Considerations. *Industrial & Engineering Chemistry Research*. **DOI:** 10.1021/acs.iecr.8b04389.
10. Pistikopoulos, E. N., Diangelakis, N. A., Oberdieck, R., Papathanasiou, M. M., Nascu, I., & Sun, M. (2015). PAROC – An integrated framework and software platform for the optimisation and advanced model-based control of process systems. *Chemical Engineering Science*, *136*, 115-138.

570300 Process Innovation and Intensification Using Building Blocks

Finding an optimal production pathway to convert raw materials into desired products is critical in chemical industry. The selection of equipment type and combinatorial nature of the equipment connectivity have been the key issue for achieving process design. Besides, discovery of novel chemical processes often requires "out-of-box" thinking and may not be achieved by simply connecting different equipment. For instance, the discovery of a feasible refrigeration cycle often requires the analysis of phase diagrams embedding rich process information for conceptual process design [1]. A novel refrigeration cycle can involve process intensification opportunities, e.g., multi-stream heat exchangers, which brings drastic improvements in term of size, sustainability, and energy efficiency [2-5]. However, discovery and identification of these chemical processes is not always intuitive and often becomes challenging while employing classic optimization-based process synthesis approaches.

Recently, building block-based process synthesis and intensification have shown promise for identifying many chemical process alternatives using fundamental phenomena and tasks [4, 6-7]. In this work, we describe a building-block based approach to enable the systematic discovery and optimization of novel process flowsheets. Thermodynamic relations from equation of state, e.g., Peng-Robinson equation of state, are described using quadratic surrogate models to reduce the computational complexity. The flowsheets will be screened among the rich connection information in the block superstructure with the surrogate thermodynamic relations. Unlike classic unit-based superstructure representation, the proposed block superstructure leverages on building blocks representing fundamental physiochemical phenomena. The block superstructure is constructed by assembling building blocks in a two-dimensional grid. These building blocks allow feed supplies and product withdrawing while connecting with each other through material and energy flow. The adjacent blocks are separated from each other through either semi-restricted boundaries for component redistribution or completely restricted boundaries for prohibiting mass flow. We will show how this building block representation yields various process configurations without *a priori* postulation of unit operations. We formulate this process innovation and intensification problem as a mixed-integer nonlinear optimization (MINLP) problem. The objective is to find innovative flowsheets with minimal total annual cost. The capability of the proposed approach will be demonstrated through several case studies on liquefied energy chain and refrigeration systems.

References:

[1] Lim, W.; Choi, K.; Moon, I. (2013). Current status and perspectives of liquefied natural gas (LNG) plant design. *Industrial & Engineering Chemistry Research*, 52, 3065–3088.

[2] Lutze, P., Gani, R., Woodley, J. M. (2010). Process intensification: a perspective on process synthesis. *Chemical Engineering Processing: Process Intensification*, 49 (6), 547–558.

[3] Demirel, S. E., Li, J., and Hasan, M. M. F., (2017). Systematic Process Intensification using Building Blocks, *Computers and Chemical Engineering*, 105, 2-38.

[4] S.E. Demirel, J. Li, M.M.F. Hasan. Systematic Process Intensification. *Current Opinion in Chemical Engineering*, 2019, https://doi.org/10.1016/j.coche.2018.12.001.

[5] Tian, Y., Demirel, S. E., Hasan, M. M. F., Pistikopoulos, E. N. (2018). An Overview of Process Systems Engineering Approaches for Process Intensification: State of the Art. *Chemical Engineering and Processing: Process Intensification*, 133, 160-210.

[6] Li, J., Demirel, S. E., & Hasan, M. F. (2018). Process synthesis using block superstructure with automated flowsheet generation and optimization. *AIChE Journal*, 64(8), 3082-3100.

[7] Li, J., Demirel, S. E., & Hasan, M. F. (2018). Process integration using block superstructure. *Industrial & Engineering Chemistry Research*, 57(12), 4377-4398.

568883 Modeling Intensification of Biomass Fast Pyrolysis through Autothermal Operation

Biomass fast pyrolysis has been identified as a viable method to produce bio-oil, a renewable alternative to petroleum-derived sources of fixed carbon. Fluidized beds are commonly used for pyrolysis due to their precise temperature control and high solid-to-solid heat transfer rates, but the largest barrier to scale-up is the difficulty of heat transfer into the pyrolysis reactor. Autothermal operation has been proven to overcome these heat transfer limitations by allowing a limited amount of oxygen within the bed, combusting a portion of the contents to supply the heat of pyrolysis. Current fluidized bed reactor models are focused on fluid-phase or heterogeneous reactions, and so are not conducive to the modeling of solid-phase reactions within a mostly inert bed.

This work presents a novel approach to the modeling of a fluidized bed reactor, focusing on the experience of small particles that make up a minority of the bed mass. Due to the decoupling of solid- and vapor-phase reactions, the group of reactive particles can be modeled as a series of open control volumes surrounded by a uniform temperature heat source. The phenomena of heat transfer, chemical reaction, density loss, and elutriation are all captured in a format applicable within commercially available process simulators. The addition of oxidation reactions to the solid and vapor phases allows the reactor model to operate autothermally. Simulation results are used to estimate the performance enhancement derived from autothermal operation.

581026 Digital Twin, Raising It out of the Realm of a Cliché

The Engineering and Industrial world are lagging manufacturing industries to truly embrace digitalization. AVEVA is accelerating the digital transformation journey of their customers and disrupting their current business processes to add value opportunities to their businesses. A common disruptive technology is the concept of Digital Twin. Through our work with our customers, we realize that believe to define a Digital Twin for the industrials we should draw a parallel to a the actual Human Twin. A twin has a shape, behavior, senses and needs to work for 100 years. The same is true for the digital twin of an industrial asset whether that is a an offshore platform, a refinery, a power generator or a ship. The Digital Twin of a plant is defined by engineering data (digital shape), engineering models (digital behavior), real-time series data (digital senses) and finally maintenance data (digital health). Once this data is connected just like a human it is can be stored in memory (historian), analyzed to make the right decisions (predictive analytics), learns from its mistakes (machine learning), asks questions (cognitive thinking) to make it more efficient (APC, real-time optimization), be mobile (smart devices) and finally visualize data in context (3D, X Reality etc...). AVEVA will present how it is connecting engineering, maintenance and operations data, people and business processes to bring the digital twin out of the realm of a Cliché.

580756 Asset Optimization Software Suites to Drive Digitalization – from Process Modeling to Optimization to Asset Optimization

Process Simulation has undergone significant transformation over the past decades driven by innovations in engineering best practices catalyzed by chemical engineering, mathematical optimization and software innovations. It has come a long way from automating the design calculations and process modeling to an environment for process optimization accelerating collaboration across the disciplines and tackling optimization across multiple dimensions simultaneously. The use of Process Simulation has also expanded from the traditional design applications to operations and planning. This digital transformation has created significant value in the industry resulting in capital and energy savings combined with better yields and capacities, increased safety and dramatic improvements in engineering productivity and production. Rapid advances in machine learning, advanced multivariate analytics, high performance mobile computing, and equipment and sensor connectivity are playing a role in enabling a whole new generation of technological solutions. The paper will cover the vision for Process Simulation including expansion from Process to Asset Optimization with first principles-based models and data-based machine learning models for process engineering, operations and performance management.

580855 A General Digital Applications Platform

Deep process knowledge captured within physics-based models is a key element for the successful digitalization of process industries. Digital applications making use of models derived from first principles are increasingly being used in industrial practice, particularly in the area of process operations. They include both open-loop applications, such as model-based soft sensing and monitoring of equipment degradation, and closed-loop ones such as Real-Time Optimization and nonlinear Model Predictive Control.

However, most non-trivial digital applications involve much more than a mathematical model being solved in a simulation or optimization mode. They often require multiple model-based calculations being scheduled over time and exchanging data with each other, while potentially being subject to occasional failures. They also involve extensive communication with external data servers, such as distributed control systems, plant historians, commercial databases and user dashboards. Moreover, the data involved in all such communications may be subject to systematic and/or random errors, and may even occasionally become unavailable.

This paper describes a recently developed general software platform for resilient and sustainable digital applications, taking account of the above considerations. The platform significantly reduces the cost and increases the reliability of development, testing, deployment and maintenance of diverse applications within a unified software architecture. Industrial examples illustrating the flexibility of the proposed design are presented.

580649 Concept of Digital Twin and Digital Thread in the Chemical & Process Industry

The vision of digital twins involves the digital twin of the product, the digital twin of process performance and the digital twin of operation.

In the R &D, stages product development and recipe transformation in with digital models can accelerate innovation. In design & engineering stages, all the disciplines involved in a plant project can generate a shared data model –saving valuable time and expenses and at the same time increasing engineering quality. The digital twin also allows the plant to be simulated prior to the critical commissioning phase, which ensures a reliable implementation. The operators can also receive targeted training before the system is commissioned. The digital twin of the operation significantly increases the reliability and availability of your plant by continuously monitoring your mechanical assets using IIoT.

The realization of this needs cross-organization collaboration. Realization of the concept needs various tools, software, hardware and collaboration platforms to work together. This talk with explore those possibilities and readiness.

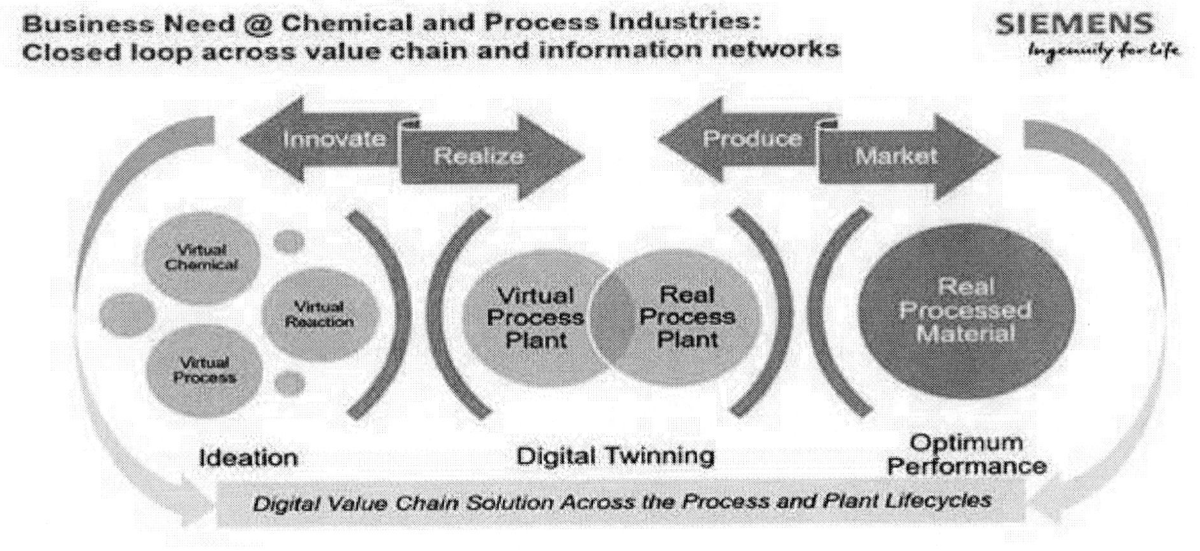

571319 Digital Twin for Aluminum Smelting

Building digital twins for manufacturing processes like aluminum smelting using first principles modeling has been considered challenging owing to the scale and complexity of such processes. A considerable mismatch is normally observed between measured data and predicted values of desired variables. This mismatch is usually because of uncertainty in the information fed to the model such as model assumptions, parameters, correlations from literature, etc. A systematic approach to identify such missing information would be to pose the model as an input-reconstruction problem. While a typical input reconstruction problem is solved using optimization-based algorithms where the objective would be to minimize the deviation between data and prediction, it may not be suitable for complex processes such as aluminum smelting. In this work we present a novel method to reconstruct missing information using a combination of first principles modeling and data analysis. The input reconstruction problem is reformulated as a set of differential algebraic equations (DAE) using measured data. Solving these DAEs help identify the missing input information and other unknown states in one shot. This method gives a complete data set (reconstructed inputs and measured data) which is used to train machine learning (ML) algorithms to build new correlations and predict unknown inputs directly in the digital twin model thereby giving more accurate predictions of desired process parameters.

579640 Transforming Abundant Data into Improved Process Reliability (Keynote Talk)

Today, processing facilities are beginning to effectively utilize the abundance of process and asset health data and information as part of an overall IIoT strategy. Although a digital transformation will require investments in technology, one cannot forget the investment in training staff to behave differently (improved) with the new timely information. It should be noted that some jobs will be replaced (e.g. manual data collection), new roles will emerge (e.g. analysts and planners), and other jobs will evolve to be more efficient. Properly trained operations are invaluable in this process, as they not only act on information from analyzed data, but can also make decisions on their own based on sensor data.

Abnormal operation and imminent asset failure can quickly be identified and corrected with properly trained staff to act on this new timely analyzed information. Big data and IIoT will continue to have a role with shortages of skilled resources, thus having the staff be more efficient and effective at their role with additional information and insight into operations.

This presentation will focus on the applications where big data and IIoT have improved process and asset reliability. Although the applications presented will come from oil refineries, the solutions can be applied to other continuous processing facilities.

579639 Going Digital in Downstream; Refinery of the Future (Keynote Talk)

The last US refinery was built in 1977, to bring these refineries into digital capability is a significant effort and we will explore actual examples and the value that refiners have received. OSIsoft participates in a project to help define the Refinery of the Future. Several examples, like a sensored pump, best uses of IIOT, 3D visualization of the plant and integrity of the asset will be shared to highlight how the downstream business is addressing these technological opportunities. We will share best practices and the progress of recognized leaders in the downstream digital transformation.

579317 Emission Monitoring and Plant Integrity Enhancement

Historically, refineries and chemical plants contribute significantly to emissions of Volatile Organic Compounds (VOCs) and Hazardous Air Pollutants (HAPs). With the increasingly stringent regulations and laws, emission management has to go beyond the traditional Continuous Emissions Monitoring System (CEMS), Leak Detection and Repair (LDAR) approach to prevent potential emission events.

To comply with the new requirement, one prescriptive approach is to enhance plant integrity. Alternative monitoring technologies, such as wireless sensing, optical sensing, facilities the engineers and operators to identify the leaks/emissions conveniently. Information technology, such as industrial 4.0, data analysis and Industrial Internet of Things (IIoT) enables the process industry to communicate and interact effectively. By integrating these technologies, plant integrity, such as predictive analytics, predictive maintenance can be conducted to improve Health, Safety, and Environment (HSE) of process industry.

This presentation describes the relevant environment regulations, emission monitoring practices, and state-of-art emission management technologies, such as predictive flare control and predictive maintenance. Cases of deployments will also be discussed in this presentation.

565604 Optimizing Real-Time Spatiotemporal Sensor Placement for Monitoring Air Pollutant for Health Impact Assessment

Air pollution exposure assessment, which involves the monitoring of pollutant species concentrations in the atmosphere, is an important step in assessing the health impact of air pollution. Currently, the air pollutants are often monitored via stationary monitoring stations. Due to the high cost of maintaining a monitoring network, the monitors can only be installed in limited number of locations. The sparse spatial coverage of monitors can lead to errors in estimating the actual exposure of pollutants in individuals. The recent advancement of cheaper sensors can help address the limitations of using stationary monitors. One approach is dynamic sensing – a new monitoring approach that adjusts the locations of portable sensors in real time to measure the dynamic changes in air quality. The key challenge in dynamic sensing is to develop algorithms to identify the optimal sensor locations in real time in the face of inherent uncertainties in the fate and transport of the air pollutants. The objective of this paper is to develop an algorithmic framework to address the challenge of sensor placement in real time. We demonstrate the capability of our algorithmic framework in a case study in Atlanta. Our real-time sensor placement optimization algorithm will allow, for the first time, the assessment of spatial-temporal variability of pollution which cannot be accomplished by a stationary monitoring station. Besides air pollution exposure assessment, our algorithm can have broader applications in improving sensor placement in drone-based monitoring for water bodies, water security networks, or advanced power systems.

562063 Energy Harvesting Sensors Using Magnetic Phase Transition

Many countries have recently become interested in cyber-physical systems (CPS) in industry and society for sustainable development. CPS equips a data acquisition function, a data storage function, and a network to transfer data. To deploy CPS, it necessary to develop overall security systems, to increase their energy efficiency and to find new energy sources. To overcome these energy-related issues, in this research we investigated the possibility of integrating data acquisition sensors with a recently developed energy harvesting system that combines the magnetic phase transition resulting from changes in temperature, and electromagnetic induction. The proposed system can provide wireless temperature or velocity sensors that directly measure electromotive forces generated by Faraday's law of induction without any additional energy input. Our proposed energy harvesting sensors have the potential to contribute significantly to the development of CPS in the near future.

579961 Mechanisms of Improvements in Virus Control By Electrocoagulation: Removal and Inactivation

Prof. Baltus and I have collaborated for over a decade on microorganism control from contaminated water supplies. We have coauthored papers on (hindered) convection and diffusion of bacteria and virus across microporous filters, but my group has been separately pursuing the idea of virus aggregation. The principal hypothesis from an environmental engineering standpoint is that virus removal is facilitated by increasing effective size using large-pored microfilters, which otherwise do not remove them. An additional advantage of coagulation is that it simultaneously removes other contaminants including natural organic matter and colloids thereby improving overall quality of our drinking water. My presentation will cover aspects of virus aggregation by externally added hydrolyzing metal-ion coagulants.

The principal objectives of this research are to elucidate underlying virus destabilization/removal mechanisms during electrocoagulation and identify possible intermediates leading to inactivation. We explore the electrochemical oxidation of sacrificial elemental aluminum and iron anodes to directly dissolve metal ion coagulant precursors that neutralize virus surface charge and upon hydrolysis sweep coagulate viruses. In all cases, the F-specific ssRNA coliphage MS2 was employed as the model virus. Atomic force microscopy was performed to directly measure adhesion forces between viruses and aluminum precipitates. Experiments were performed on synthetic buffered water at low and high pH as well as using saline background solutions to electrolytically produce free chlorine. Low chlorine concentrations combined with virus shielding and aggregation within flocs resulted in very slow disinfection rates in solutions with high salinity.

Inactivation mechanisms were pursued via attenuated total reflectance – Fourier transform infrared spectroscopy and quantitative reverse transcription polymerase chain reaction. Evidence for oxidative transformations of capsid proteins including formation of oxyacids, aldehydes, and ketones and RNA damage kinetics will be discussed. Finally, recent results on electrochemical iron inactivation and coagulation will be presented.

565782 Fabrication of Catalytic Water Treatment Membranes with Phosphorene-Based Pore Fillers

Photocatalytic membrane processes generate oxygen-reactive radicals under irradiation by light that could degrade pollutants in feed solutions and reduce fouling. Phosphorene is a two-dimensional material made by exfoliating black phosphorus, a monotypic van der Waals crystal. Unlike other common two-dimensional materials, under light irradiation, phosphorene can generate reactive oxygen species (singlet oxygen, hydroxyl radicals and superoxide radicals). Therefore, it shows promise as a photocatalyst. The goal of this study was to incorporate phosphorene into membranes made from a polymer blend of sulfonated poly ether ether ketone and polysulfone, then test these membranes in different media to examine their stability in various environment and study the method of adhesion of the nanomaterial in the membrane. In this research, phosphorene was first fabricated by the liquid exfoliation of black phosphorus. Experiments were done in an acidic, basic, neutral and in a common solvent media. Various membranes that composed of different wt.% of phosphorene were casted and the rate of leaching of the phosphorene nanoparticle was studied. Results obtained showed enhanced flux recovery for the phosphorene modified membranes. Fluorescent studies also revealed lower coverage of methylene blue on these membranes as compared to the unmodified membranes. Mechanism of photocatalysis will be investigated and reaction equations proposed.

574425 Tangential Flow for Analyte Capture (TFAC): A Method for the Purification of Nanoparticle Biomarkers Enabled By Ultrathin Silicon Membranes

Extracellular vesicles (EVs) are lipid vesicles ranging from 30 – 1000 nm in diameter secreted by cells to serve a variety of functions. Certain members of the EV family, such as exosomes, are thought to be important for intercellular communication. Exosomes contain cell specific biomarkers (i.e. miRNA, protein) that are useful for point-of-care (POC) diagnostics. A variety of techniques are used to isolate EVs, including ultracentrifugation, affinity-based capture, precipitation and filtration, but these often require large sample volumes (50 – 300 mL) which are not feasible for a POCsystem. Additionally, these isolations are often contaminated with protein, leading to difficulty in the identification of biomarkers. Therefore, an alternative method of isolating EVs from small volumes is requiredfor POC. Silicon nanomembranes are ultrathin (50-100nm) nanoporous membranes discovered during the rapid crystallization of pure silicon films. These membranes are manufactured with a variety of pore diameters (~ 5 nm – 100 nm) and have high porosities (20% – 40%) and exceptional permeabilities. These membranes are manufactured in a 'chip' format that is 5.4 mm x 5.4 mm, making them perfect for small volume applications such as microfluidic filtration. In this work, we present a novel exosome capture methodforsmall volume (~ 1 mL)samples.We callthis methodtangential flow analyte capture (TFAC). After demonstrating the capture ofexosomes from undiluted serum, wepresenta computational model for TFACthat is validated with 60 nm gold nanoparticleswith and without the addition of serum albumin (2mg/ml). We usethe model to define optimalcaptureconditions, and thendemonstrate the capture of > 75% of exosomes from defined protein/exosome sampleand compare these results to capture from raw urine.

579962 Advanced Membrane Materials for Environmental Liquid Separations

Our group focuses on fundamental understanding of the relationships between formation, structure, material properties, and material performance in advanced membranes for applications in environmentally relevant processes.

Surface and groundwaters have increasing concentrations of salts from agriculture and water softening activities—increasing the need for inland brackish water desalination and concentrate management. During inland brackish groundwater desalination, disposal of the concentrate can often constrain the design and overall feasibility. Reverse osmosis (RO) is typically used to achieve 75 to 85% recovery of the feed water and limited by sparingly soluble salts (e.g., calcium sulfate) or minerals (e.g., silica) and the overall concentration of the feed. Pervaporation is a membrane process that separates mixable liquids by a combination of liquid permeation through and vapor evaporation from a dense semi-permeable membrane. It uses hydrophilic dense nonporous (polymer membranes) or microporous (inorganic membranes), in which the driving forces are vapor pressure of the main permeating species and, indirectly, temperature. In pervaporation there is no limit on the salinity of the feed water. We will discuss our results to incorporate direct solar light to enhance the pervaporation process for desalination as well as our development of new membrane for pervaporation.

Liquid fossil fuels are a significant portion of the annual energy consumption of the United States. Butanol is a second generation biofuel whose production does not compete with food production and butanol is a suitable substitute for liquid fossil fuels. Using a condensation-cure method we have synthesized free-standing polydimethylsiloxane/zeolite-imidazolate framework-71 (PDMS/ZIF-71) mixed-matrix membranes with ZIF-71 loadings of up to 40%. Our membranes have the highest reported selectivity for pervaporation recovery of ethanol and butanol from dilute aqueous solutions (selectivity of 5.64 ± 0.15 for 2 wt% 1-butanol in water and a selectivity of 0.81 ± 0.04 for 2 wt% ethanol in water). Tensile testing indicates that the Young's modulus of the membranes increased with higher ZIF-71 loadings. We also demonstrate that the organic linkers found in ZIF-71 are incompatible with addition-cured PDMS films.

576709 Highly Selective Ultrafiltration – Exploiting Non-Steric Interactions for Bioprocessing

Ultrafiltration is a well established method for protein concentration and buffer exchange. Nearly all current biotherapeutics are formulated using ultrafiltration / diafiltration processes. However, recent advances in our understanding of membrane transport phenomena has led to exciting opportunities for the development of highly selective ultrafiltration processes for bioseparations. This talk will review these developments and how they have been used to develop membrane systems for the purification of conjugated vaccines and nucleic acid-based gene therapy agents. These results clearly demonstrate the significant potential of using ultrafiltration membranes for the commercial production of next generation biotherapeutics.

577326 Additive Manufacturing of Thin Film Composite Membranes

Our recent work on printing polyamide membranes demonstrated the ability to make desalination membranes of exceptional smoothness and controllable thickness. The approach required the electrospray deposition of an amine monomer and an acid chloride crosslinker onto a substrates where they react to form an aromatic polyamide. The technique allows for exceptional control of the reaction and makes it advantageous over conventional interfacial polymerization. Conventional interfacial polymerization is a relatively uncontrollable process and has few options for independently tailoring selectivity and permeance. We can adapt this approach to making membranes of novel polymers as well, many of which cannot be formed easily into thin film or composite structures with high permeance *and* high selectivity. We use this approach with several emergent polymer systems materials to demonstrate the value of this approach.

573504 Emerging Organ Models and Organ Printing for Regenerative Medicine

Engineered materials that integrate advances in polymer chemistry, nanotechnology, and biological sciences have the potential to create powerful medical therapies. Our group aims to engineer tissue regenerative therapeutics using water-containing polymer networks called hydrogels that can regulate cell behavior. Specifically, we have developed photo-crosslinkable hybrid hydrogels that combine natural biomolecules with nanoparticles to regulate the chemical, biological, mechanical and electrical properties of gels. These functional scaffolds induce the differentiation of stem cells to desired cell types and direct the formation of vascularized heart or bone tissues. Since tissue function is highly dependent on architecture, we have also used microfabrication methods, such as microfluidics, photolithography, bioprinting, and molding, to regulate the architecture of these materials. We have employed these strategies to generate miniaturized tissues. To create tissue complexity, we have also developed directed assembly techniques to compile small tissue modules into larger constructs. It is anticipated that such approaches will lead to the development of next-generation regenerative therapeutics and biomedical devices.

571530 3D Layered Nano- Micro- Fibre Printing Techniques for Biological and Electronic Applications

Building two-dimensional (2D) and three-dimensional (3D) fibrous structures in the micro-nano scale will offer exciting prospects for numerous applications spanning from sensors, to energy storage and tissue engineering scaffolds. Existing fibre spinning techniques have been tailored for producing micro-to-nano scale fibers, but can be restrictive in the design of macroscopic fiber architectures in 3D space, and the choice of fibre materials. To address these technological gaps, we report two new fibre spinning techniques: first, 3D-LEP which combines low voltage electrospinning (LEP) and additive manufacturing to pattern suspended fiber layers in multiple tiers and designable orientations; and secondly, an efficient inflight fluidic fibre deposition (IFD) process to produce core-shell conducting fibres. Using these techniques, we demonstrate biological applications utilising fiber topography to guide the assembly of cellular aggregates in a 3D culture context; and optoelectronic applications showing unconventional transparent, suspended fibre-array electronics and the construction of layered 3D electrical architectures.

570135 Putting 3D Biofabrication to the Use of Tissue Model Fabrication

Microphysiological systems are microfluidic three-dimensional miniature human tissue and organ models that recapitulate the important biological and physiological parameters of their *in vivo* counterparts. They have recently emerged as a viable platform for personalized medicine and drug screening. These biomimetic microtissues are anticipated to supplement the conventional planar, static cell cultures, and to bridge the gaps between the current pre-clinical animal models and the human body. In addition, multiple microtissues may be channeled together through the microfluidics in a similar manner they arrange *in vivo*, providing the capacity to analyze interactions among these models. In this talk, I will specifically discuss our recent efforts on developing a series of bioprinting strategies including sacrificial bioprinting, microfluidic bioprinting, and multi-material bioprinting, along with various cytocompatible bioink formulations, for the fabrication of biomimetic 3D tissue models. These platform technologies, when combined with bioreactors and bioanalysis, will likely provide new opportunities in constructing functional tissue and disease models for drug development, therapeutics screening, and precision medicine.

570597 3D Printing Functional Voxels: Fate and Transport of Quantum Dots Under Capillary Flow

Nanotechnology is a technological branch that focuses on nanomaterials with dimensions smaller than 100 nm. Nanotechnology has and continues to impact many industries: petroleum, materials, cosmetics, and pharmaceutics, etc[1]. More than 1,300 consumer products based on nanotechnology have entered the marketplace[2]. Some common nanomaterials include: silica, fullerol, alumoxane, carbon nanotubes, metal oxide nanoparticles, nanocrystals, and quantum dots.

For three dimensional printing (3DP) companies, the ability to integrate these nanomaterials into additive manufacturing is a desirable outcome. By incorporating these nanomaterials into 3DP, it allows for an even broader platform of materials. Expanding the materials portfolio empowers the consumer, enabling those individuals the potential to develop and engineer novel materials to exploit new functional properties.

Understanding the transport and fate of these nano-sized materials is vital as these materials may eventually contaminate the subsurface environment and become a threat to the watertable[3]. Concern arise as evidence proposes that some nanomaterials can negatively impact human health and the environment[2]. A fundamental understanding of the fate of nanomaterials, specifically nanoparticles, is imperative for both applications as well as environmental and health impact.

This presentation proposes a novel non-invasive technique with high temporal and spatial resolution for imaging nanoparticle transport and retention in polymeric powders used for 3D printing. Engineering on the voxel-scale is a multi-physics process involving the infiltration, transport, adsorption, and retention of engineered nanoparticles (ENP's) suspended in a carrier solvent. Fundamentally, solute transport and retention in porous media involves macroscopic fluid dynamics, mesoscale advection-dispersion processes, and competing microscale physiochemical forces and interactions (e.g. DLVO forces) that occur at the interfaces between the solute particles and the porous media[3–5].

1. Zhang, T. *et al.* Investigation of Nanoparticle Adsorption During Transport in Porous Media. *SPE J.* **20**, 667–677 (2015).
2. May, R., Akbariyeh, S. & Li, Y. Pore-Scale Quantification of Nanoparticle Transport in Saturated Porous Media Using Laser Scanning Cytometry. *Environ. Sci. Technol.* **(in review**, 6–12 (2012).
3. Torkzaban, S., Bradford, S. A., Wan, J., Tokunaga, T. & Masoudih, A. Release of quantum dot nanoparticles in porous media: Role of cation exchange and aging time. *Environ. Sci. Technol.* **47**, 11528–11536 (2013).
4. Bradford, S. A. & Torkzaban, S. Colloid Transport and Retention in Unsaturated Porous Media: A Review of Interface-, Collector-, and Pore-Scale Processes and Models. *Vadose Zo. J.* **7**, 667 (2008).
5. Goldberg, E. *et al.* What Factors Determine the Retention Behavior of Engineered Nanomaterials in Saturated Porous Media? (2017). doi:10.1021/acs.est.6b05217

570866 Radio Frequency and Microwave Heating of Preceramic Polymers and Its Application in Silicon Carbide Processing

Here, we demonstrate microwave and noncontact Radio Frequency heating methods for mold free processing of silicon carbides (SiC) via preceramic polymer (polycarbosilane) route using carbon nanotube susceptors as fillers. This results in volumetric heating of polycarbosilanes and has applications in 3D printing of SiC and fiber processing.

We show that polycarbosilanes have minimum response to RF and addition of electromagnetic susceptors like MWCNT results in rapid heating and curing of these polymers. We perform dielectric characterizations on the polycarbosilane-MWCNT composites at room temperature. We were able to achieve 95 % extent of curing for polycarbosilane-MWCNT composite using RF in range of 76-78 MHz at 100 W. Microwave pyrolysis of polycarbosilane-MWCNT composites results in formation of cubic SiC structure. We used noncontact RF heating method for 3D printing ink composed of polycarbosilane and SiC nanoparticles with iterative deposition and curing to form a cross-linked structure. We also report for the first time change in dielectric constant and ac conductivity of the 3D printing ink as it heats and transforms while curing. We have studied rheological properties of the ink and its shear thinning behavior. We have also shown that RF/microwave heating results in molecular heating and avoids puddle formation while curing polycarbosilane fibers to SiC fibers.

577247 Complex Multi-Layered Microparticle Laden Fixed Hydrogel Wound Dressing Fabricated Via 3D Printing for the Treatment of Thermal Burns

While burns are a common injury in society today, the damage that burns – especially partial-thickness and full-thickness burns – can cause to bodily function and appearance cannot be underestimated. While skin grafting remains as the gold standard for the treatment of partial and full thickness burns, autologous skin grafting suffers from the prerequisite of needing sufficient healthy skin from the patient while the risk of rejection of donor skin by the patient's body is a concern in allogeneic skin grafting.

As either an alternative or a complement to skin grafting, wound dressings can be applied to provide a conducive environment that promotes healing and minimizes microbial infection. With the emergence of 3D printing technologies, the creation of complex 3D structures from a myriad of materials can be achieved inexpensively and efficiently. In this study, we propose the development of multi-layered fixed hydrogel wound dressings using 3D printing for the treatment of thermal burns. To further aid the wound healing capabilities of this wound dressing, microparticles encapsulating relevant drugs such as topical antibiotics and analgesics are incorporated into the wound dressing. 3D printing strategies used in the production of these hydrogel wound dressings along with their structure and function will be discussed.

Through this study, we hope to present novel strategies in which the strengths of 3D printing can be leveraged to fabricate complex hydrogel wound dressing structures with better efficacy in the healing of thermal burn wounds.

563909 Modeling Rosette Nanotubes for Applications in Water Filtration, Biosensing and Drug Delivery

Rosette nanotubes (RNTs) are biocompatible supramolecular nanostructures that are formed via self-assembly of building blocks of Watson-Crick DNA-inspired guanine-cytosine (G C) motifs. Similar to the double helix of DNA, hydrogen bonding between the individual motifs lead them to assemble into rings, called rosettes.[1,2] A combination of π-π interactions between the rings and hydrophobic effects lead the rosettes to self-assemble into nanotubes, where the rings either stack or assemble into helical coils. Because of its biocompatibility, RNTs have attracted attention for drug delivery and biological applications, such as encapsulating dexamethasone to enhance cell growth in bones.[3] RNTs can also have potential applications in nanopore-based sensors and water filtration membranes. Here we report on our current studies that involve (1) modeling the interactions of RNTs with cyclodextrins and water, and (2) developing a Martini-based[4] coarse-grained model for the RNTs.

1. Fenniri, B. L. Deng, and A. E. Ribbe. *J. Am. Chem. Soc.* 2002, **124**, 11064-11072.
2. Fenniri, B. L. Deng, A. E. Ribbe, K. Hallenga, J. Jacob, and P. Thiyagarajan, *Proc. Natl. Acad. Sci. USA* 2002, **99**, 6487-6492.
3. Borzsonyi, R. Johnson, A. Myles, J. Cho, T. Yamazaki, R. Beingessner, A. Kovalenko, and H. Fenniri. *Chem. Commun.*, 2010, **46**, 6527-6529.
4. Uusitalo, H. Ingólfsson, P. Akhshi, D. Tieleman, and S. Marrink. *J. Chem. Theory Comput.*, 2015, **11**, 3932–3945.

580662 Enhancing Tractability of Model Predictive Control-Assisted Online Data Collection

A goal of next-generation manufacturing is to utilize control to enhance efficiency and improve profits [1]. To integrate economic optimization and process control, an optimization-based control design termed economic model predictive control (EMPC) has been studied [2, 3], which uses a stage cost that is not required to have its minimum at a process steady-state. This type of control design is of interest, for example, for potentially enabling next-generation manufacturing systems to account for the impacts of time-varying costs on profitability. More accurate process models may aid in the development of this model-based control design, which raises the question of whether it is possible to obtain physics-based models automatically from process data. Though techniques have been developed which seek to fit potentially more physics-based models to data using techniques such as symbolic [4] or sparse [5] regression, there is still a need for additional techniques which can postulate model structures based on the trends in process data, so that parameters can be subsequently fit to such models. One idea for achieving this is to utilize EMPC to guide the process state over time to take values which correspond to information that is meaningful for discriminating between various model structures that might be postulated [6]. However, control designs which have these capabilities also must be examined for tractability in design and computation time.

Motivated by the above, this work focuses on methods for enhancing the tractability of EMPC-assisted online data collection by exploring a variety of techniques which can increase practicality of the methods. For example, we explore a distributed control framework for driving the process state between operating conditions in state-space when multiple Lyapunov-based EMPC (LEMPC) [7] designs are available that can be selected between at a given sampling time, which guide the process state toward different desired data. We also explore methods for reducing the computation time required for mechanisms which trigger a decision to gather desired data rather than operate the process in a standard fashion and suggest techniques for developing stability region estimates for LEMPC on-line. A chemical process example is used to demonstrate the proposed methods in terms of closed-loop stability, feasibility, and computation time reduction.

References:

[1] J. Davis, T. Edgar, J. Porter, J. Bernaden and M. Sarli. Smart manufacturing, manufacturing intelligence and demand-dynamic performance. *Computers & Chemical Engineering*, 47:145-156, 2012.

[2] M. Ellis, H. Durand, and P. D. Christofides. A tutorial review of economic model predictive control methods. *Journal of Process Control*, 24:1156–1178, 2014.

[3] D. Angeli, R. Amrit and J. B. Rawlings. On average performance and stability of economic model predictive control. *IEEE Transactions on Automatic Control*, 57:1615-1626, 2012.

[4] M. Schmidt and H. Lipson. Distilling free-form natural laws from experimental data.*Science*, 324:81-85, 2009.

[5] S. L. Brunton, J. L. Proctor and J. N. Kutz. Discovering governing equations from data by sparse identification of nonlinear dynamical systems. *PNAS*, 113:3932-3937, 2016.

[6] L. Giuliani and H. Durand. Data-based nonlinear model identification in economic model predictive control. *Smart and Sustainable Manufacturing Systems*, 2:61-109, 2018.

[7] M. Heidarinejad, J. Liu and P. D. Christofides. Economic model predictive control of nonlinear process systems using Lyapunov techniques, *AIChE Journal*, 58:855-870, 2012.

580663 Model-Guided Design of a Solar-Assisted Water Desalination Process through Membrane Distillation

Membrane distillation (MD) is an emerging technology for desalination of seawater and brackish water. MD is a thermal process in which the desalination occurs by diffusion of water vapor across a hydrophobic porous membrane in contact with hot saline/feed (e.g., 50-80 ^{o}C) and cold freshwater/distillate streams. In MD, the driving force for transmembrane vapor permeation is the difference between the saturation vapor pressure of water across the membrane. The operating pressure and temperature in MD process are very mild; this is the main advantages of MD over other commercialized water desalination technologies such as reverse osmosis (RO) and mechanical vapor compression (MVC). Hence, the MD process can be coupled with affordable energy sources such as solar energy or waste heat. Nonetheless, MD has not been fully commercialized mainly due to lack of proper design of material and devices (e.g., membrane modules) that makes the process relatively energy intensive. For this reason, we applied a model-guided approach to design a high-performance integrated solar-assisted hollow fiber membrane module. To do so, a three-dimensional (3D) multi-physics model of a hollow fiber membrane module was developed, and the effect of operating and design parameters on the module performance was investigated. The permeate flux and thermal efficiency of the system were considered as the characteristic parameters of the module. The simulation results indicated that the permeate flux for the module can be enhanced by 54% if the module is designed properly. In the next step, the results of our multi-physic simulations were used to write energy and mass balance around a solar-MD system. A set of ordinary differential equations (ODEs) was developed, and the influence of module and membrane parameters on the performance of the system was taken into consideration. Both design and processing conditions were searched to find an operational condition that allows for operating closer to the thermodynamic limits. To design and fabricate a Solar-MD system, the importance of a few key design parameters: module properties such as length, packing density, membrane properties, processing conditions such as flow rates, and system disturbance such as fluctuation in solar irradiance were evaluated.

580659 Multiscale Modeling of Pulp Fiber Length in Batch Pulping Process

During the last two decades, several efforts have been made to recover fibers from used paper and increase the recycling rate, thus achieving a recycling rate of 64.7 percent (paper and paperboard product) in the United States [1]. However, paper products cannot be recycled permanently as the fiber length will be shortened after every recycling process and eventually become too short to make papers (generally, recycle up to six times). Since the fiber length governs the interfiber bonding capability which is associated with tensile and burst strengths of paper, the shortened fibers cannot be the source of high quality paper but can be used for newspaper or egg cartons [2-4]. Accordingly, a precise modeling of fiber length is required to elucidate its evolution during the pulping process which may increase the paper recycling efficiency. However, the existing mathematical models are limited in describing the evolution of pulp microscopic properties such as fiber length, cell wall thickness (CWT) and porosity [5-7].

Motivated by this limitation, we developed a multiscale model that is capable of describing both macroscopic and microscopic phenomena of the batch pulping process. Specifically, by integrating the most widely used mathematical model for pulping process (i.e., the Purdue model) [8] and a multiscale coarse-grained Monte Carlo (CGMC) algorithm [9], the evolution of both the Kappa number (i.e., residual lignin content in pulp), fiber length and CWT are captured accurately. The mass continuity and thermal energy balance equations are adopted from a modified "extended Purdue model", and microscopic events such as dissolution of solid molecules are executed by the kinetic Monte Carlo algorithm [10-11]. Since the fiber length (*mm*) and CWT ( ) have different length scales, multiscale CGMC algorithm is implemented to calculate the probability density function of both the fine-grid and coarse-grid scales by separately operating two lengths and two time scales. Additionally, as a novel aspect of the proposed model, practical considerations like the water vessels and cell wall ultrastructure (e.g., chemical composition of cell wall layers) of wood chip were considered in the simulation lattice to precisely describe the realistic evolution of aforementioned microscopic properties. Furthermore, several other important fiber morphological parameter measurements can be estimated by the proposed multiscale model. For example, an increase in fines is generally observed during the recycling operation which can improve strength by improved interfiber bonding capability. However, only indirect determination methods are available, providing a potential application to our proposed model as a soft sensor for the properties that are not measured real-time.

References

- Ali I. Study of the mechanical behavior of recycled fibers. Applications to papers and paperboards. *D. thesis,* University of Grenoble, Grenoble, France, **2006.**
- James LM.; Rajai HA. Strength loss in recycled fibers and methods of restoration. *Res. Soc. Symp. Proc.*, **1992**, *266*, 218-228.
- Johansson A. Correlations between fibre properties and paper properties. *Master's thesis,* KTH Royal Institute of Technology, Stockholm, Sweden,
- Facada MJ. Influence of Kraft paper quality on the performance of an industrial paper impregnation process. *D. thesis*, Universidade Tecnica de Lisboa, Lisboa, Portugal, **2015.**
- Funkquist J. Grey-box identification of a continuous digester – a distributed-parameter process. *Control Eng. Pract.*, **1997**, *5*, 919-930.
- Wisnewski PA.; Doyle FJ. Fundamental continuous pulp-digester model for simulation and control. *AIChE J.*, **1997**, *43*, 3175-3192.
- Smith C. Studies of the mathematical modeling simulation and control of the operation of Kamyr continuous digester for the Kraft process. *D. thesis,* Purdue University, West Lafayette, IN, **1974**.
- Bhartiya S.; Dufour P.; Doyle FJ. Fundamental thermal-hydraulic pulp digester model with grade transition. *AIChE J.*, **2003**, *49*, 411-425.
- Chatterjee A.; Vlachos DG. Multiscale spatial Monte Carlo simulations: Multigriding, computational singular perturbation, and hierarchical stochastic closures. J. Chem. Phys., **2006**, *124*, 064110.
- Kwon JSI.; Nayhouse M.; Christofides PD.; Orkoulas G. Protein crystal shape and size control in

batch crystallization: Comparing model predictive control with conventional operating policies. *Eng. Chem. Res.*, **2014**, *53*, 5002-5014.
- Choi HK.; Kwon JSI. Multiscale modeling and control of Kappa umber and porosity in a batch-type pulp digester. *AIChE J.* (In press), **2019**.

576998 FDM of ABS-Nanoclay Composites

FDM of ABS-Nanoclay Composites

Burcin Ikizer[1] and Azadeh Farahanchi[2]

[1]Chemical Engineering, UMass Lowell, Lowell, MA, United States

[2]Dynisco, 38 Forge Parkway, Franklin MA 02038

Materials used in additive manufacturing should display a favorable balance of mechanical and rheological properties. Acrylonitrile-butadiene-styrene (ABS) terpolymer is one of the most commonly used plastics in Fused Deposition Modeling (FDM) due to its ease of processability. ABS by itself is considered to be at the lower end of the spectrum of engineering plastics due to its low impact resistance and toughness. Its mechanical and chemical properties can be enhanced through the use of fillers. However, currently the use of filler-reinforced thermoplastics is limited in 3D printing due to the poor weld strength of the printed materials. This study aims to investigate mechanical properties of printed parts made from pure ABS as well as ABS-nanoclay composites with a FDM 3D printer. Built in stresses of the printed parts will also be studied.

577941 Experimental Modeling of a Nonlinear Hybrid Dynamic System

An experimental multivariable pressure tank system was built inspired by previous experimental systems for liquid level modeling and control applications. The dynamic response of the pressure system exhibits significant nonlinear characteristics. A hybrid dynamic nonlinear model has been established to capture the switching behavior for flow through valves. The expression for flow through a valve can be modeled as either normal flow or choked flow when the pressure drop across the valve is high. Additionally, it was found that the control valve flow rate expression can be further refined by including additional hybrid switching behavior to implement an adjustable saturation behavior.

The valving of this experimental four-tank system can be operated to establish a number of possible SISO and MIMO configurations. The current work considers a relatively simple two-tanks in series configuration. The resulting nonlinear model includes expressions for flow across three different valves. Parameter estimation routines provide estimates for the various model parameters. Optimal switching conditions and model parameters have been determined from experimental data. A comparison of different types of models ranging from basic linear to full hybrid nonlinear will be presented.

Data from the system can be examined using phase plane analysis. The hybrid dynamic model presents transitions between different hybrid dynamic modes in the phase plane. For the experimental system, seven different discrete modes are reachable depending on operation of the system. Using the model, an input sequence may be developed so that numerous transitions between different hybrid states can be excited. This allows for improved empirical modeling of the experimental system by creating a data series for identification purposes which includes numerous hybrid mode transitions.

580656 A Workflow for the Fabrication of Innovative Crystallizers Using 3D-Printing

3D printing offers the unique possibility to fabricate customized process equipment at reduced cost and without compromising on the fabrication time. The high potential of 3D printing as a manufacturing technology can especially be exploited for the fabrication of process equipment with a complex geometry and no moving components. These advantages of 3D printing have been exploited for the fabrication of micro and meso-scale reactors.[1,2]In particular, all of the four major 3D printing techniques (i.e., stereolithography[3], selective laser sintering[4], polyjet printing[5]and fused deposition molding[6]) have been used in the fabrication of reactors. Despite the wide utilization in reaction engineering, 3D printing has found limited applications in the fabrication of equipment for separation processes, including crystallization. Crystallization is an important separation and purification technology in various industries such as the fine chemical and pharmaceutical industry. Recently, we have developed a 3D-printed airlift crystallizer for batch protein crystallization[7]. 3D printing proved to be useful due to the need for customization of the complex geometry and the absence of moving parts in such crystallizer. Furthermore, crystallization processes often operate under mild conditions, which allows for different types of materials to be used. Other types of crystallizers may also be fabricated efficiently with 3D printing. For example, tubular crystallizers with mixing internals[8,9]would be another class of innovative crystallizers that could utilize the advantages offered by 3D printing. Different types of crystallizers may require different 3D printing techniques. Furthermore, in general, a workflow that guides decision making when using 3D printing to fabricate prototype crystallizers can facilitate faster adoption by industry. However, such workflows or advisory documents are currently lacking.

The objective of this work is to present a workflow that will guide a user through the different steps and decisions that are needed to use 3D printing for the physical realization of a crystallizer. The major steps in the development of the crystallizer include, selection of crystallizer type, sizing of the equipment based on the process requirements, an optional step for computational analysis (e.g., CFD simulations), selection of an appropriate 3D printing technique, detailed 3D drawing of the equipment, selection of a 3D printer and the material, fabrication of the equipment and finally testing. An important step in the development of a crystallizer is the selection of an appropriate 3D printing technique. The key considerations involved in the selection of the 3D printing technique include the required accuracy, build resolution, ease of the removal of the unused material and support structures, if any, and the ability to print stable leak-proof parts. An additional advantage is that the fabrication speed and lower fabrication cost with 3D printing makes it possible to have multiple iterations of the above-mentioned workflow in a relatively short duration, which is difficult to achieve with traditional manufacturing methods. A workflow detailing the above-mentioned steps will be presented. Finally, the application of the workflow will be demonstrated for the case of an airlift crystallizer. The possibility to fabricate innovative process equipment rapidly and at a lower cost makes 3D printing an attractive manufacturing technology in situations where early learning of the process is important but time to market has to be as fast as possible or when the attrition rate of new product concepts is high, such as in the pharmaceutical industry.

Acknowledgement: The work described in this abstract was supported by a grant from the Research Grants Council of the Hong Kong Special Administrative Region, China, Project No. 16242916.

References

(1) Kitson, P. J.; Rosnes, M. H.; Sans, V.; Dragone, V.; Cronin, L. Configurable 3D-Printed Millifluidic and Microfluidic “lab on a Chip” Reactionware Devices. *Lab Chip*__2012__, *12*(18), 3267–3271.

(2) Lin, C. G.; Zhou, W.; Xiong, X. T.; Xuan, W.; Kitson, P. J.; Long, D. L.; Chen, W.; Song, Y. F.; Cronin, L. Digital Control of Multistep Hydrothermal Synthesis by Using 3D Printed Reactionware for the Synthesis of Metal–Organic Frameworks. *Angew. Chemie - Int. Ed.*__2018__, *57*(51), 16958–16962.

(3) Okafor, O.; Weilhard, A.; Fernandes, J. A.; Karjalainen, E.; Goodridge, R.; Sans, V. Advanced Reactor Engineering with 3D Printing for the Continuous-Flow Synthesis of Silver Nanoparticles. *React. Chem. Eng.***2017**, *2*(2), 129–136.

(4) Maier, M. C.; Lebl, R.; Sulzer, P.; Lechner, J.; Mayr, T.; Zadravec, M.; Slama, E.; Pfanner, S.; Schmölzer, C.; Pöchlauer, P.; et al. Development of Customized 3D Printed Stainless Steel Reactors with Inline Oxygen Sensors for Aerobic Oxidation of Grignard Reagents in Continuous Flow. *React. Chem. Eng.***2019**,*4*(2), 393–401.

(5) Lee, K. G.; Park, K. J.; Seok, S.; Shin, S.; Kim, D. H.; Park, J. Y.; Heo, Y. S.; Lee, S.-J.; Lee, T. J. 3D Printed Modules for Integrated Microfluidic Devices. *RSC Adv.***2014**, *4*, 32876–32880.

(6) Rossi, S.; Porta, R.; Brenna, D.; Puglisi, A.; Benaglia, M. Stereoselective Catalytic Synthesis of Active Pharmaceutical Ingredients in Homemade 3D-Printed Mesoreactors. *Angew. Chemie - Int. Ed.***2017**, *56*(15), 4290–4294.

(7) Mathew Thomas, K.; Lakerveld, R. A 3D-Printed Airlift Crystallizer for Protein Crystallization. *Submited***2019**.

(8) Alvarez, A. J.; Myerson, A. S. Continuous Plug Flow Crystallization of Pharmaceutical Compounds. *Cryst. Growth Des.***2010**, *10*(5), 2219–2228.

(9) Yang, H.; Peczulis, P.; Inguva, P.; Li, X.; Heng, J. Y. Y. Continuous Protein Crystallisation Platform and Process: Case of Lysozyme. *Chem. Eng. Res. Des.***2018**, *136*, 529–535.

570666 Thermal Effects on Capillary Flow in Polymeric Powders

3D Printing (3DP) is a new, exciting, innovative, and disruptive technology. For 3DP companies, the ability to integrate nanomaterials into additive manufacturing is a desirable outcome. By incorporating these nanomaterials into 3DP, it allows for an even broader platform of materials. Expanding the materials portfolio empowers the consumer, enabling those individuals the potential to develop and engineer novel materials to exploit new functional properties.

Powder bed technology has the technological ability to push the materials portfolio. The powder bed technologies have the ability to introduce nanoparticles to the powder bed via the dispensed fluidic agent. To successfully implement a variety of nanoparticles into a variety of powder beds, fundamental approaches to understand capillary flow in polymeric powders at elevated temperatures need to be developed.

This presentation discusses a novel technique to measure relative contact angles of powder as a function of surface wetting at a variety of temperatures. The preliminary results for polycarbonate, polystyrene, and polyamide-12 have been promising. This technique may be beneficial to individuals or groups interested in in the 3D printing technologies, as well as those investigating flow through porous media or packed beds at elevated temperatures. The experiments reach temperatures of 65°C. This technique is a viable method to probe interfacial phenomena at varying temperatures.

580650 An Advanced 3D Computational Model for Designing Next Generation Drug Carriers

Multi-compartment microcapsules are promising transport vehicles for next generation of smart drug delivery carriers. Several diseases and disorders such as HIV require simultaneous delivery of multiple active pharmaceutical ingredients. In such cases, the active ingredients must be delivered to the targeted area of the human body and released simultaneously. Thus, multiple smaller microcapsules including different active ingredients are encapsulated inside a mother microcapsule. The present work utilizes a state-of-the-art advanced computational methodology (membrane spectral boundary element method) that accurately predicts multi-compartment microcapsules deformation and motion in vascular channels. The developed three-dimensional model provides an enhanced understanding on the complex hydrodynamic behavior of these drug carriers. The interfacial interactions of inner capsules with the mother capsule in confined geometries are studied, and insights for optimization and advance manufacturing of next generation of smart drug carriers are presented.

580661 3D Printing and Robotic Filling of Multi-Compartment Capsular Devices for Oral Drug Delivery

Therapies involving the oral administration of multiple drugs that either are mutually incompatible or interact in the gastrointestinal tract would benefit from the availability of multi-compartment capsular devices. These systems would simplify the dosing schedule and enhance patient compliance. Particularly, our technology enables efficient fabrication of fixed-dose combination products in small, personalized batches.

In the research work here presented, fabrication of capsule parts and relevant filling have been described. The devices under development comprise a modular structure resulting in the presence of multiple independent compartments. Each of them could be filled with various drugs and with a range of strengths/formulations of the same active component. Moreover, the release behavior of a single compartment would mainly depend on the composition and thickness of its walls. Therefore, by adjusting these parameters, the release performance of each compartment can be modified without affecting the properties of the joint compartments, thus yielding multiple release kinetics in a single unit.

By enabling the separation of the development of the capsule shell from the drug formulations conveyed, these systems would also ease time to market of new products.

In this respect, two-compartment capsular devices composed of three parts were initially developed. More into detail, capsule shells comprising 400 µm-thick and 800 µm-thick compartments were fabricated using 3D printing by fused deposition modeling (FDM), starting from hydroxyl propyl cellulose (Klucel™ LF). This required the development of *i)* an extrusion plant for manufacturing of filament suitable for 3D printing and of *ii)* an industrial-grade FDM equipment, both compliant with the strict requirements of the field of interest. In both cases, identification of the critical operating parameters that would affect the quality attributes of the final product were performed, together with the definition of suitable in-process controls. We assessed the stability of the starting polymer after hot processing, ruling out possible contaminations and formation of any hazardous products. The results obtained confirmed that extruded filaments and printed prototypes meet the specifications set.

Once filled and assembled, the two-compartment capsular devices showed the expected two-pulse release behavior, consistent with characteristics of the starting material employed. Furthermore, dimensional and release characteristics of multi-compartment capsules prepared by FDM turned out analogous to those produced by injection molding. Both represent interesting manufacturing techniques, for either rapid and inexpensive fabrication of small on-demand batches or relevant large-scale production.

A purposely developed robotic system, based on automated filling modules working simultaneously and in parallel, was employed for filling each compartment with specific formulations. Cross-contamination was avoided by enclosing each filling station in a dedicated clean room and by employing a robotic arm to move the capsular devices form one filling station to the next. The performance of the filling stations was evaluated during preliminarily filling trials with powders having diverse particle size and bulk density.

This level of personalization required innovative, non-destructive controls in order to guarantee the quality of the products in a cost-effective manner. Quality control machines able to check the thickness and weight of the capsule parts during the process have been developed. Spectroscopic techniques are also going to be implemented to confirm the identity of formulations to be used for filling.

580657 Material Extrusion Based Additive Manufacturing with Blends of Polypropylene and Hydrocarbon Resins

Filament material extrusion (MatEx) is one of the most commonly used techniques in additive manufacturing due to its operational flexibility, easy maintenance, low energy demands, low material wastage and low cost. However, in spite of notable development in the MatEx process, there are currently only a handful of materials that can be printed consistently using this technique. Isotactic polypropylene (PP), a popular thermoplastic material, undergoes rapid crystallization that leads to significant trapped residual stress when processed using MatEx, resulting in poor adhesion to the printing platform, poor geometric tolerance and mechanical performance. In this work, the effects of low molecular weight hydrocarbon resins on the thermal behavior, crystallization, and morphology of their blends with PP are studied. The addition of the resins to the pure PP matrix lowered the crystallization temperature of PP from 121.8°C to 116.0°C, which enables additional diffusion during the solidification process. Polarized optical microscopy demonstrates the differences in crystalline morphology, which impacted the structure at the interlayer boundaries between deposited roads in particular. The combination of modifications in crystallization rate, time, and morphology significantly affect the z-axis adhesion and residual stress state, which directly controls the mechanical properties and part warpage. Tensile bars were printed in two different orientations to analyze the mechanical performance and study part anisotropy. The maximum tensile stress for one of the printed 80/20 blends is 32.5 MPa which is higher than that of the printed PP part (26.8 MPa). The improvement in the tensile strength is due to a combination of changes in crystallinity, morphology, and improved interlayer adhesion during printing. The parts were annealed post-printing to improve polymer chain diffusion across the layers thereby improving interlayer adhesion and resulting in tensile modulus and strength values in excess of 90% of injection molded (IM) PP parts.

580658 Effect of Temperature on the Morphology of Extruded Strand in Fused Deposition Modeling: A CFD Study

Fused Deposition Modeling (FDM) is an extrusion-based additive manufacturing technique. In FDM process, 3D object is built by extrusion and deposition of polymer melt, layer by layer, on a substrate plate. Although there have been several numerical modeling studies regarding FDM, most of these studies focus solely on either inside or outside of the nozzle, leading to incomplete evaluation of the process. In this study, isothermal and non-isothermal simulations of the process, considering both inside and outside of the nozzle, have been performed. For this purpose, a user-defined module has been developed for OpenFOAM CFD package. Numerical simulation results for shape of an extruded strand have been validated by comparing with experimental results. Moreover, the effects of geometrical and process parameters on the deposition of the layer and the surface roughness of final products are studied.

580651 Process Intensification in Solid Dosage Formulation By Hot Melt Extrusion

The pharmaceutical community has realized the need to shift from batch to continuous manufacturing to minimize cost and increase efficiency. Traditional batch methods for solid dosage formulation (up to date, the most convenient and industrially relevant drug delivery for peroral administration) often involve multiple costly and time consuming powder handling steps such as milling, wet or dry granulation, drying, sieving, and tableting. Common efforts to transition from batch to continuous manufacturing often focus on mere integration of traditional batch processes into a continuous production line. However, continuous drug product manufacturing allows to decrease and/or combine process and handling steps by innovative integration of excipients and active pharmaceutical ingredients (APIs) via process intensification. Hot melt extrusion (HME) represents such a process intensification technology for continuous solid dosage formulation, especially when coupled with injection molding or *3D* printing to produce well-defined shapes and sizes of drug products in a single step. When contrasted with traditional solid dosage formulation processes, HME offers various advantages including; (1) reduced processing steps, (2) reduced processing time, (3) reduced space, energy and carbon foot-print, (4) solvent-free, and (5) increased manufacturing capacity and efficiency due to continuous operation. However, to date, polymorphic stability of an API is thought to be a prerequisite to employ HME. Polymorphism, a phenomenon that enables molecules to exhibit multiple crystalline phases, is one of the most scrutinized critical quality attributes of solid drug products. Polymorphism is estimated to occur in 80% of APIs, affecting properties of the solid state and, therefore, the quality and efficacy of the final drug product. The unsubstantiated notion that undesired polymorphic phase transformations might occur during the HME process limits the application of this process intensification technique to a narrow set of APIs that are both thermally stable and monomorphic (~20%). The present study showes that despite the process intensification by HME polymorphic control can be achieved, similar to traditional continuous solid dosage formulation efforts, simply by understanding and controlling critical process parameters that accompany a HME process (e.g. temperature, pressure, drug-polymer composition, polymer properties, and residence time). Drug products were analyzed by powder-x-ray diffraction and Raman spectroscopy and compared to conventional products.

572846 Invited: Sensor Applications in Bioengineering: Detection of Small Molecules within Complex and Three-Dimensional Biological Milieu

Bioengineering applications in regenerative medicine and single-cell analysis present challenging microenvironments for diagnostic and real-time detection. We have developed and applied sensors to detect small molecules in neuronal, cardiovascular, blood and biofilm systems. In particular, we have aimed to establish easy-to-use sensor platforms that do not interfere with biological processes. These sensors are reagentless and can be deployed in the presence of complex macromolecular solutions that readily foul, degrade or block many traditional sensing approaches. Herein, we describe our efforts towards detection of neuroactive molecules in the brain, hypoxia gradients in three-dimensional cellular scaffolds, and subcellular oxygen concentrations during electrophysiological and chemo-mechanical single cell experiments.

575136 Tissue-like Neurochemical Sensors Operating in Brain and Gut

Normal neurotransmission relies on a precise and delicate balance of chemical signaling mediated by the release of variety neurotransmitters. Among them, the monoamines including dopamine and serotonin are uniquely involved in the regulation of cognitive processes such as emotion, arousal and memory. Therefore, it is of great interest in both basic research and clinical studies to monitor the fluctuation of monoamine in both brain and peripheral nervous systems. As a very soft and complex biomaterial, the human brain is under constant motion and deformation because of different physiological dynamics, while GI tract is a series of soft organs with inherent motility, joined in a long and twisting tube from the mouth to the anus. As a result, rigid neurochemical sensors can cause inflammation to the brain tissue, and neurochemical sensing in awake animals have historically been impossible to traditional rigid and fragile devices.

Softness and stretchability in new-generation neuroengineering tools are essential to minimize the perturbation of motions inherent in biological systems and maximize the physiological fidelity. Here we develop a new neuroengineering platform, based on soft materials, that could allow access to the brain and entire enteric nervous system for neurochemical sensing. The new neurochemical sensor is based on graphene-elastomer composite, which offers chronic, sensitive, and selective sensing of dopamine and serotonin in the brain. Ex vivo studies shows that the soft sensor could accommodate the peristalsis movement of the intestine and continuously monitor the serotonin dynamics via thermal treatment. In vivo experiments also show that the sensor could be implanted in the colon of mice to examine the drug-stimulated serotonin response. The new neuroengineering platform described here can provide unprecedented insights on the microbiota–gut–brain communication in the context of mental disorders and therapeutic interventions, and can lead to new therapies for psychiatric disorders through GI tract without invasive surgery to the brain.

570825 Cell-Based Electrochemical Sensors for Environmental Pollutant Detection

Endocrine disrupting compounds are ubiquitous; they are found in many products intended for common use, including pesticides, plastics, and even some medications. Yet, these chemicals are implicated in health problems ranging from diabetes to cancer. We have developed a cell-based electrochemical strategy to rapidly and sensitively quantify these compounds. Our strategy measures the biological activity of chemically dissimilar xenoestrogens by harnessing the specificity of a human estrogen receptor. This receptor is expressed on the surface of *E. coli* as a component of an electrochemical sandwich assay. Our sensors require no specialized skills to implement and have enabled very sensitive detection (sub-ppb levels) of environmental pollutants. We have successfully detected compounds in from complex solutions, such as baby formula, and have measured estrogenic activity from unknown compounds leeched from a BPA-free baby bottle.

577305 Biomimetic Composites As Microfabricated, Flexible, Degradable Electrochemical Sensors

Biomimetic composites as microfabricated, flexible, degradable electrochemical sensors

- M. Xu, S. Pradhan, R.K. Pal, V.K. Yadavalli

Department of Chemical and Life Science Engineering, Virginia Commonwealth University, Richmond, VA

Biomimetic composites of naturally derived and synthetic polymers provide exciting opportunities to develop mechanically flexible, physiologically compliant, and degradable bioelectronics systems. In prior work, we have shown that naturally derived silk proteins and conducting polymers (specifically poly(3,4-ethylenedioxythiophene) :poly(styrene sulfonate) (PEDOT:PSS)) can form functional bioinks for sensing and other electrochemical applications. These inks combine the mechanical strength, biocompatibility and degradability of silk proteins, with the conductivity and chemical functionalization of the PEDOT:PSS. The fabrication and characterization of photopatternable, water-based conductive inks comprising PEDOT:PSS and synthesized photoreactive silk proteins was earlier shown [1] The presence of photoreactive groups permits a fully aqueous photolithographic strategy to form conductive microelectrodes on both rigid substrates as well as flexible silk films [2].

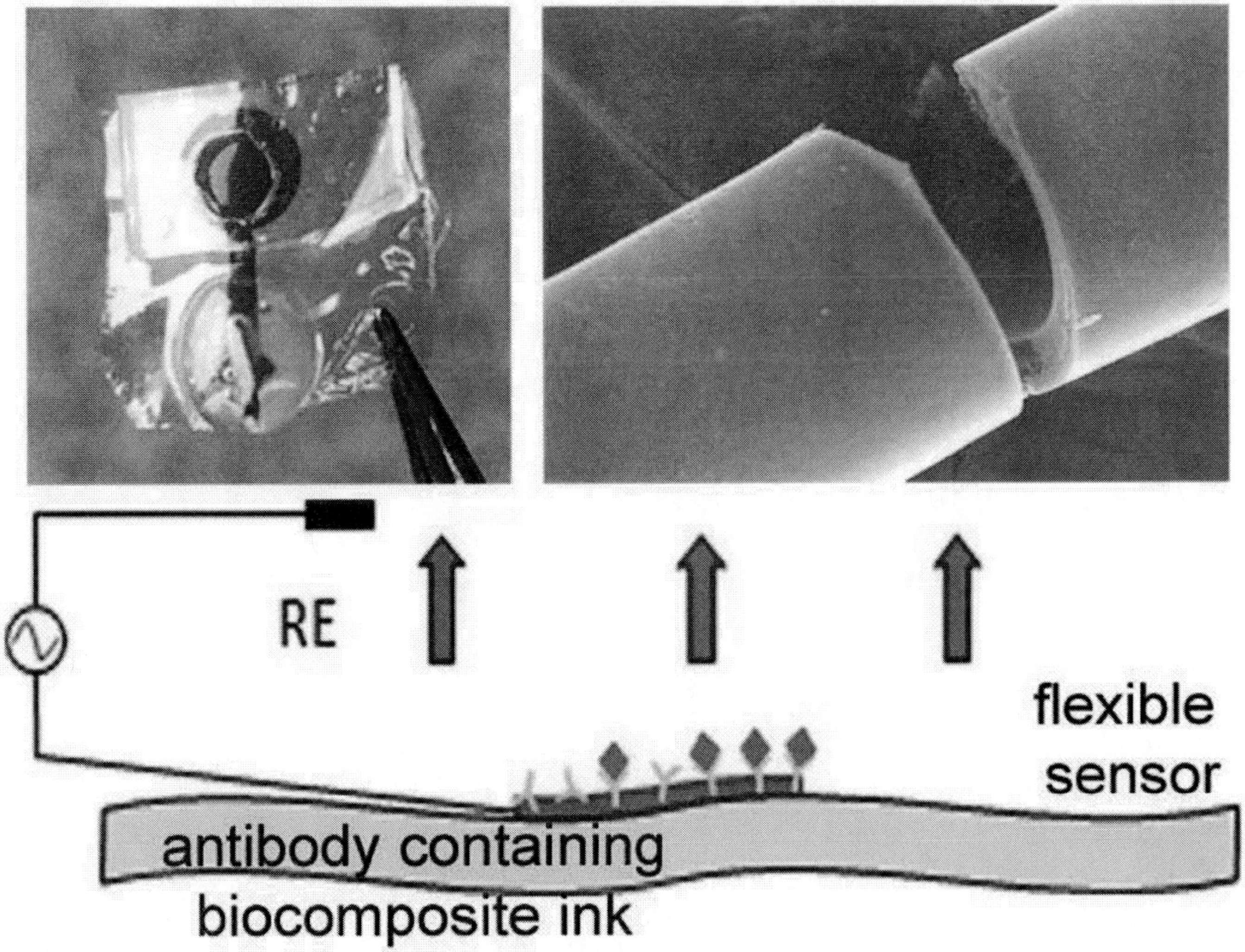

Here we will present the latest developments using this technology, showing how these devices can function for flexible bioelectronics applications in the form of microelectrodes for ultrasensitive detection, as well as components such as microsupercapacitors.[3, 4] We show how electroactive biomolecules such as neurotransmitters can be detected sensitively, while non-electroactive biomolecules such as glucose and glutamic acid, can be detected by encapsulating specific enzymes. Due to the room temperature, aqueous fabrication process, we can immobilize antibodies in these sensors resulting in flexible immunosensors for targets such as vascular endothelial growth factor (VEGF) for the first time [3]. The electroactivity of conductive ink can be improved by the addition of small amounts dopants. Using these doped composites, we further demonstrate flexible energy storage devices due to the capacitive nature of the biomaterial. The presence of silk proteins as the matrix of the composite makes it completely biodegradable and biocompatible, potentially resulting in transient or implantable devices. This platform can also be used for the specific control of cell culture in flexible sheets. The mechanical, biochemical and electrochemical characterization of the composite and its microfabrication are discussed. By virtue of a range of versatile properties, utility as bio-sensors, opto-electronic devices and flexible energy storage systems are envisioned.

References:

[1] "Photolithographic micropatterning of conducting polymers on flexible silk matrices" - RK Pal, AA Farghaly, MM Collinson, SC Kundu, VK Yadavalli, *Advanced Materials*, 28(7), 1406-1412, 2016

[2] "Conducting polymer-silk biocomposites for flexible and biodegradable electrochemical sensors" - RK Pal, AA Farghaly, MM Collinson, SC Kundu, VK Yadavalli, *Biosensors & Bioelectronics*, 81, 294-302, 2016

[3] "Flexible biosensors for the impedimetric detection of protein targets using silk-conductive polymer biocomposites" – M Xu, VK Yadavalli, *ACS Sensors*, in press, 2019

[4] "Fabrication of Flexible, Fully Organic, Degradable Energy Storage Devices Using Silk Proteins" - RK Pal, SC Kundu, VK Yadavalli, *ACS Applied Materials & Interfaces* 10 (11), 9620-9628, 2018

569635 Nanopore Based Highly Sensitive and Accurate Detection of HIV

Nanopore Based High Performance Point-of-Care Biosensor for Acute HIV Detection

Xiaojun Wei, Chang Liu

Human immunodeficiency virus (HIV) is the causative agent of acquired immuno-deficiency syndrome (AIDS). There are over 1.8 million infections HIV occur annually worldwide, and 30%¨C50% are attributed to recently infected individuals that are unaware of their disease state. Rapid diagnosis of acute HIV (within 3 weeks of infection) may reduce new cases by 10%¨C25%. The fourth-generation rapid diagnostic test for HIV is currently employed worldwide as the frontline screening test. This Enzyme-linked Immunosorbent Assay (ELISA) based test is designed to detect HIV-1 p24 antigen in addition to HIV antibodies to achieve early detection. However, several recent field tests have indicated unacceptably low diagnostic sensitivity (0%) of acute HIV-1.

Herein, a nanopore biosensing platform with single-molecule level sensing and analysis capacity has been developed to address the unmet need for rapid and accurate diagnosis of acute HIV to reduce transmission. In our approach, p24 antigens isolated from plasma samples were cleaved into several specific peptide sequences, which can be recognized and quantified by their signature signals during translocation through the nanopore. A portable QuantiPep-HIV prototype suitable for point-of-care settings was developed and tested with serum samples from HIV-infected and uninfected individuals to demonstrate the feasibility and to evaluate the diagnostic performance. Our results showed that QuantiPep-HIV can improve the diagnostic sensitivity and specificity of acute HIV.

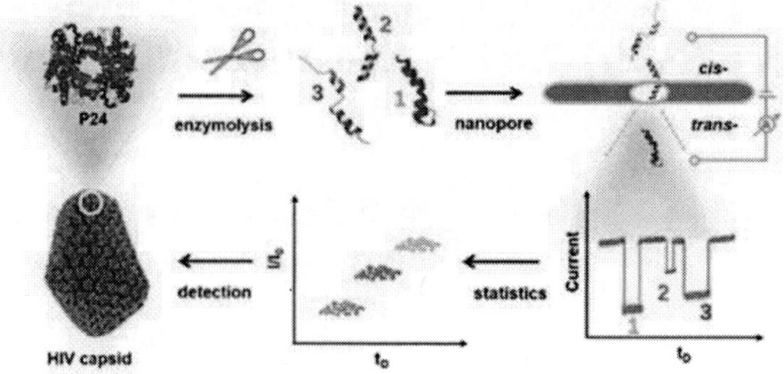

Figure. Schematic diagram for detecting peptide fragments from p24 antigens by nanopore

582378 Wireless, Soft, Skin-integrated Microfluidic Systems for Electrochemical/Colorimetric Biosensing and Capture of Sweat

Human eccrine sweat is a rich composition of nearly 60 known biochemicals including metabolites, electrolytes, vitamins, amino acids, exogeneous drugs, and small proteins. Advances in sensing technologies, conformal electronics, and materials science lay the foundation for recent demonstrations of wearable sweat sensors. Unfortunately, majority of the devices rely either on active, battery-powered electronics for electrochemical detection or passive, colorimetric chemistries for visual readout. Complex construction, large size/weight and possibility of battery leakage represent disadvantages of the former; semi-quantitative operation and limited scope of measurable biomarkers are drawbacks of the latter. Additionally, present wearable sensors can detect only a handful of biomarkers present in the sweat. Quantification of the remaining chemicals still requires sweat collection using adsorbent pads followed by off-site chemical analysis. Such a route provides average concentration of sweat analytes and restricts the acquisition of their temporal concentrations. This prohibits the understanding of the relation between dynamic variations in sweat chemistry and human physiology. In this talk, I will describe an unusual collection of enabling materials, sensing technologies, advanced microfluidics, stimulus dependent biocompatible power systems, galvanic time-tracking schemes, and wireless electronics to address these issues. Specifically, I will describe my strategies that include soft microfluidics, fuel cells, trigger-activated biocompatible batteries, colorimetrics and near field communication technologies that result in new classes of wireless, skin-integrated devices for detailed analysis of human sweat.

574301 The Electrochemical Impedance Response of a Continuous Glucose Monitor

The Gerischer impedance has been observed in systems with an electrochemical reaction coupled with a homogeneous chemical reaction. Under the assumptions of linear homogeneous reaction, equal diffusion coefficients and a semi-infinite condition, the Gerischer impedance can be solved analytically [1, 2]. However, with non-ideal electrochemical systems, where coupled nonlinear homogeneous reactions and various mass-transfer properties make exact analytical resolution impossible, a more detailed analysis is required.

In the present work, a mathematical model for the impedance response of glucose-oxidase based electrochemical biosensors has been developed [3]. The complicated coupling between thirteen homogeneous reactions and three heterogeneous reactions has been considered in the model. The homogeneous reactions included anomerization between α-D-glucose and β-D-glucose and four reversible enzymatic catalytic reactions transforming β-D-glucose and oxygen into gluconic acid and hydrogen peroxide, pH-dependent enzymatic activity and a biological buffer system. The electroactive hydrogen peroxide was considered to be reversibly oxidized or reduced at the electrode.

The mathematical model is solved numerically by using the finite-difference method and Newman's BAND algorithm [4]. The model demonstrates how the coupled non-linear homogeneous reactions effect the diffusion impedance, which has broadened the scope of the Gerischer impedance. The model can be used to explore the influence of various system parameters on limiting current, reaction profiles, and diffusion impedance. The system parameters, including interstitial glucose concentration, oxygen concentration, active enzyme concentration, diffusion coefficients, reaction rate constants and layer thickness, are related to various sensor working conditions such as body sugar level, inflammation, sensor degradation and sensor design.

References

1. H. Gerischer, "Wechselstrompolarisation Von Elektroden Mit Einem Potentialbes- timmenden Schritt Beim Gleichgewichtspotential," *Zeitschrift fur Physikalische Chemie*, **198** (1951) 286–313.
2. M. E. Orazem and B. Tribollet, *Electrochemical Impedance Spectroscopy* (John Wiley & Sons, Hoboken, NJ, 2017), 2nd edition, p. 279-293.
3. M. Gao, M. S. Hazelbaker, R. Kong, and M. E. Orazem, "Mathematical Model for the Electrochemical Impedance Response of a Continuous Glucose Monitor," *Electrochimica Acta*, **275**(2018),119-132
4. J. S. Newman and K. E. Thomas-Alyea, *Electrochemical Systems* (John Wiley & Sons, Hoboken, NJ, 2004), 3rd edition.

Acknowledgement

The support of Medtronic Diabetes (Northridge, CA) and Andrea Varsavsky, program monitor, is gratefully acknowledged.

580905 Introductory Remarks for the Industry 4.0 and Digital Transformation Session

The distinguished industrial speakers will be introduced by the session chair.

579773 Leading the Industry 4.0 Journey in the Manufacturing Sector with Cross-Industry Collaboration and AI-Powered Innovation

While manufacturing represents 16% of global GDP, manufacturing industries account for 64% of global R&D spend. Yet innovation is long overdue since the manufacturing sector has experienced a decade of productivity stagnation and demand fragmentation. Organizations that have taken Industry 4.0 innovation to scale beyond the pilot phase have experienced unprecedented increases in efficiency with minimal displacement of workers. However, most companies appear to be stuck in "pilot purgatory". Widespread adoption of artificial intelligence (AI) and allied Industry 4.0 technologies at scale, through the combined efforts of companies and governments, can lead to a sizeable increase in global wealth creation, benefiting people throughout society.

However, there is potential for worker displacement if changes are not handled properly. Strengthened collaboration among stakeholders is needed to understand unexpected consequences and manage the transition.

Leaders in digital manufacturing and globalization 4.0 exhibit all the essential characteristics of the Industry 4.0 revolution. Moreover, they confirm the hypothesis that they have the potential to generate new economic value, driven by improvements in the full spectrum of production value drivers: resource productivity and efficiency, agility and responsiveness, speed to market and customization to meet customer needs. Transforming existing production systems, innovating value chains and building new business models with disruptive potential for established businesses can create value.

579424 Real-Time Analytics for IIoT

Digital transformation, connected enterprise, and IIOT are various names that refer to an increasing interest by the manufacturing industry to make better use of the data they generate. A key driver behind this heightened interest is the expectation that the business will benefit from mining actionable information out of that data.

The prevailing discourse today equates successful digital transformation with big data analytics (often delivered via a cloud-centric infrastructure). This talk will argue that, at least in industrial applications, there are a significant subset of problems for which a cloud-centered big data approach is not necessary.

This talk will present a family of embedded analytics capabilities that are designed to extract actionable information out of streaming operation data. A brief description of each capability along with the results from real-world applications of the technology will be presented.

579314 Process Optimization in the Age of Industry 4.0

Industry 4.0 represents a new trend of automation and data exchange in manufacturing technologies and fosters the concept of a "smart factory". Within this framework cyber-physical systems monitor physical processes, create a virtual copy of the physical world and make decentralized decisions. This talk provides an overview of modern optimization optimization tools and solution strategies within this distributed decision-making process. In contrast to current engineering tools which are based on simulation models and expand to optimization, we describe the development of more advanced decision-making tools that start with an optimization model and flesh out the details of the process. In particular, we highlight the development of equation oriented optimization platforms with fast solvers, advanced modeling environments and efficient interfaces between them. Moreover, we extend these features to NLP sensitivity, nonsmooth models and data-driven models. Finally, we explore the application of these tools for distributed optimization, optimization under uncertainty and the integration of design, operations and control.

580904 Value-Driven Digitalization Initiatives at Dow

Largely, the process industry has matured beyond the initial frenzy around automated reports, dashboards, and the embedding of statistics in software products, and has firmly moved toward building out infrastructure and decision-support systems to improve safety and maximize value.

Industry megatrends around M&A for narrowed domain focus, commoditization, low capital intensity, geo-political uncertainty, and emphasis on sustainability and customer experience require nimble and responsive supply chains, systems analysis for understanding trade-offs, and maximization of value from asset footprints. Further, with the emergence of AI platforms, the time has never been more ripe for practitioners to help an allegedly sluggish and slow-to-adopt industry to embed digital tools as a core part of their strategy.

A lot of emphasis has rightly been on external visibility of product portfolios to customers, as well as on internal decisions around what, when, and where to produce product---consequently involving IT, customer service/marketing, and supply chain functions. Although there continues to be significant opportunity in this space, AI technology and analytics has become increasingly relevant to R&D groups in the development of new products and innovative processes; and to maintenance and operations organizations in predicting and responding to process events.

Further, most digitalization projects fall in one of three areas of increasing implementation complexity: (1) technology that enables improved decision-making in a local environment that can be then cloned in different environments with small tweaks; (2) technology that is specific to a particular domain, business, class of products, or process; or (3) technology that spans across multiple assets and allows consolidated decision-making at scale. These three categories differ in the effort needed to standardize, scale, and get buy-in from stakeholders. These pose slightly different challenges, and the way organizations structure support organizations, empower employees, and manage programs is critical to the successful deployment of these systems and in effecting culture change.

Here, we provide specific examples where Manufacturing 4.0-related projects have enabled product, process, maintenance, and supply chain innovation at Dow, including recent success stories of deploying process intensification at scale. We also cover what Dow has done to intensify its organizations through three key capabilities---the Digital Marketplace, the Digital Operations Center, and the Digital Fulfillment Center, where domain experts and engineers sit alongside software architects, UX designers, and database experts to deliver solutions effectively. Further, we discuss how communities of practice around analytics have helped us align analytics professionals across different functions to develop a coherent strategy for the company around, for example, deployment best practices, building data pipelines, and digital twins. External partnerships with universities, consultants, and vendors do play a key role in technology transfer and adoption, and balancing the development of in-house talent with the right external support is important as well.

Through these stories, we hope to illustrate how an organization can plan a steady and firm climb out of the trough of disillusionment with digital technologies into the plateau of productivity.

580906 Q&A Panel Discussion

The distinguished speakers will hold a Q&A session with the audience, moderated by the session chairs.

580371 A Case Study for the Use of Conventional and Modular Construction Methods in Building an Intensified Chemical Plant

Early efforts in offsite construction grew out of project needs such as building chemical plants in remote locations lacking the infrastructure for supporting plant construction. More recently, mainstream chemical companies are using offsite construction to reduce capital costs and project timelines, putting capital to work faster and lowering the risk of entering new markets. Engineering and construction firms have realized shorter construction timelines, lower construction costs and improved work safety using offsite construction to overcome labor shortages, enhance labor productivity and improve quality of work. However, most chemical manufacturers have not realized numbering-up strategies for managing capacity and most engineering and construction firms have stopped short of economies of mass production during off-site construction. Process intensification strategies to process development lend themselves to modular construction as intensified processes are more capable of fitting within modular form factors. Within the RAPID institute, the modular construction of intensified plants is finding its way forward within many distributed chemical processing markets where capacity is scaled up by numbering up, leading to offsite construction efficiencies associated with building the same module multiple times. This talk will provide insights from a case study showing the key drivers associated with the use of modular offsite construction for a distributed chemical processing application. The differences in capital equipment costs for the intensified chemical plant built using conventional "stick-built" construction methods versus modular offsite construction methods will be contrasted. The methodology described within this talk will be used to collect lessons learned from other modular chemical process intensification efforts, ongoing within the RAPID institute, for advancing specialty chemical production.

573225 Modular Ethane-to-Liquids Via Low-Temperature Chemical Looping Oxidative Dehydrogenation of Ethane

North American shale gas contains large amounts of natural gas liquids (NGLs). One NGL, ethane, poses particular transport challenges for producers. Unlike propane and butane, ethane is difficult to liquefy for transport, and ethane pipeline capacity is limited. This leads to frequent ethane "rejection" by reinjection into natural gas streams. Even rejection of ethane as fuel presents challenges due to interstate natural gas pipeline specifications. A modular scale process that can convert ethane into valuable and easily transportable products at geographically isolated gas processing facilities would capture significant value from this underutilized domestic resource. Ethylene production via steam cracking of ethane, which is the major industrial use of ethane, is not well suited to modular applications. The necessary high temperatures (>1000 °C), extensive heat exchange and recovery, and cryogenic separations demand large-scale operation to be economical.

Here we present a modular Ethane-To-Liquid fuel (m-ETL) system that is enabled by low-temperature (≤750 °C) redox catalysts for chemical looping-oxidative dehydrogenation (CL-ODH). In this approach, ethane is partially oxidized to ethylene and water by utilizing the lattice oxygen of a redox catalyst in a packed bed reactor. After the redox catalyst is depleted, the reactor is purged followed by the introduction of air to both regenerate the lattice oxygen and provide heat for the net reaction. By bundling multiple packed beds together in parallel, heat can be integrated between reaction steps. The ethylene-rich product stream is then compressed and oligomerized to light liquid olefins suitable for use as gasoline. Gaseous hydrocarbon byproducts, which are recycled to the reactor with a slip stream that is burned in an engine to provide power for the system. The ability to use number-up manufacturing for the CL-ODH bundles coupled with the modular scale of these bundles make them well-suited for low-cost distributed implementation. Process simulations based upon experimental redox catalysts highlight the potential productivity of a modular system. The economic potential of and challenges to successful implementation of the technology are also explored.

580362 On Demand Treatment of Wastewater Using 3D-Prined Membrane

Over the last 40 years, membrane-based separation processes have evolved from laboratory scale applications to a multibillion-dollar technology widely employed by the food processing, chemical production, and pharmaceutical industries. Fundamentally, these separation processes exploit the selective permeability of a membrane substrate to segregate multi-component mixtures, either via an imposed pressure (e.g., microfiltration and reverse osmosis) or concentration (e.g., dialysis and forward osmosis) gradient. While traditional membrane-based technologies represent an energy-efficient alternative to more robust methodologies (e.g., fractional distillation), such processes are inherently limited to a single separation mechanism (i.e., a single selectivity). Mixtures featuring three or more components, or those with multiple components and phases, cannot currently be separated in a single processing step. The single membrane selectivity also results from the present-day fabrication techniques, which limits the construction to a single material and pore distribution.

To combat the deficiencies in the current paradigm, a novel 3D-printed membrane capable of simultaneous multicomponent and multiphase separation is undergoing development in our lab. Taking advantage of innovations in additive manufacturing and 3D printing technology, the established fabrication technique permits localized control over membrane composition, as well as pore size and 3D topography so that single membrane will have multiple selectivity. Currently, two distinct separation mechanisms are being investigated in our lab, using a combination of experimental and computational methods. The first one is a gravity-fed separation of a multiphase oil-water mixture. With water-assisted separation or/and 3D topography design, we have shown that the wettability of the membrane can be tailored to enhance the separation efficiency. The second one is a diffusion-based separation of a miscible multicomponent mixture through a supported ionic-liquid membrane (SILM). With the optimization of both ionic liquid, which serves as the extraction solvent, and membrane structure/chemistry, the miscible liquid mixtures are separately successfully. Moving forward, we plan to combine the two mechanisms in single 3D-printed membrane and demonstrate the separation of multi-component and multi-phase liquid mixtures.

578403 Selecting Membrane Distillation Configuration

Membrane distillation is an emerging thermally-driven membrane technology proposed for the separation of salts and other non-volatile compounds from wastewater streams. The process offers a unique solution for the treatment of water streams with high total dissolved solids. There are four configurations proposed for MD applications: Direct Contact MD, Vacuum MD, Sweeping Gas MD, and Air Gap MD. Each of these configurations has its own advantages and disadvantages, and none of them has been commercialized, yet. Studying some of these MD configurations (e.g. DCMD) have been limited to laboratory-scale, while some (e.g. SGMD) have been investigated for pilot scale applications. Comparative analysis of these configurations has been hindered by lack of infrastructure and technology limitations. Here we present a comprehensive analysis of MD configurations based on process-level Aspen simulations. Four MD configurations has been simulated using Aspen Custom Modeler. Several performance metrics such as energy efficiency, capital and operating costs, flux delivery, and potential heat integration technologies are investigated, and results will be presented. We will report the energy requirements of each configuration to treat one cubic meter of water ($\$/m^3$).

580631 Deploying Intensified, Automated, Mobile, Operable, and Novel Designs "Diamond" for Treating Shale Gas Wastewater

The DIAMOND project is aimed at developing integrated design and operating approaches of modular systems that can be deployed for treating flowback and produced water resulting from shale gas production. Because of the highly distributed nature and variable characteristics of shale-gas wastewater (SGWW), there is a unique opportunity to deploy modular systems. There is also a major challenge in developing tailored designs for each source of wastewater. An integrated theoretical-experimental project is proposed to: (1) Assess, screen and integrate commercially-viable conventional and emerging technologies for wastewater treatment, (2) Develop computer-aided modeling, design, operation, scheduling, and costing approaches for non-recurring engineering needed to deploy the SGWW treatment systems, and (3) Demonstrate proof-of-concept via applications to a broad range of SGWW samples. A combination of systems engineering approaches and experimental/pilot-scale work will be used to generate commercially viable design and operational strategies with significant impact on intensification metrics.

A dynamic model will be developed to describe the flow rate and TDS concentration of flowback and produced water from fractured wells. These models will be validated against experimental data available from the literature. Based on this, a computer tool will be developed to run detailed complex models according to modular/intensified process alternatives, obtained from superstructure optimization. An engineering study will be performed to provide techno-economic and controllability analysis.

Based on the characterization of available samples, a combination of experimental/pilot scale work will be conducted. USCWT technology uses electromagnetic fields and electrocoagulation to clean shale gas wastewater by removing, oxidizing, or precipitating contaminants. Bench and pilot scale units will be used to demonstrate this technology with waters from gas fields studied in this project.

Membrane distillation laboratory experiments will be carried out using simulated and real SGWW to identify membranes that are resistant to very high TDS concentrations. Simulated samples will be created by preparing brines with total dissolved solids (TDS) concentration between 50,000 ppm and 150,000 ppm. Parameters such as turbidity and total suspended solids (TSS), as well as SGWW-contained minerals ions as well as heavy metals, will be included in the study. Both lab-scale and pilot-scale experiments will be conducted to validate operating parameters and system performance.

Ionic liquids (ILs) have the potential to extract salts, as well as some organics, from SGWW. The key is to prevent back contamination of the wastewater with the IL. Manufacture, test and model polymeric membranes will be designed to allow transport of the SGWW contaminants through the membrane, but prevent the IL from entering the cleaned aqueous phase.

The technical and economic feasibility of membrane distillation (MD) and counterflow reverse osmosis (CFRO) for treatment of shale gas wastewater (SGWW) will be evaluated. Lab and pilot scale data will be used to inform and develop optimization-based technoeconomic models for SGWW treatment using MD and CFRO technologies.

In conclusion, all these processes will be used to build the framework for assessing, screening and integrating commercially-viable conventional and emerging technologies for wastewater treatment.

580967 Approaches to Knowledge Management and Data Transformations in Pharmaceutical Process Invention

Over the past decade, the increased molecular complexity of small molecule drugs has resulted in synthetic complexity. However, traditional unit operations (reactions, separations, crystallization, filtration, etc.) continue to be the basis for the manufacture of small molecule drug substances even as innovations in organic synthetic techniques such as new catalysts and coupling reagents are implemented to improve sustainability and efficiency. Chemical engineers are well-poised to support chemical route invention, process invention and process characterization activities to develop robust commercial manufacturing processes through utilization of limited data to quantify and metricize competing attributes of proposed synthesis approaches and process designs. Specifically, methods to understand and quantify process risks, process capabilities, and expected failure rates have the potential to speed development activities. Addressing these challenges requires innovations in tools supporting data generation and curation, and new approaches to metrics and modeling. This talk will cover some current work in this knowledge space and its potential to drive innovation in pharmaceutical development in the future.

581329 Leading the Way to Pharma 4.0

Manufacturing processes and plant & supply chain scheduling are becoming increasingly more complex. Pharma 4.0 is helping transform plants into smart factories. This is being done through improvements in sensor technology, real-time monitoring, advanced process control, continuous process verification, predictive maintenance, digital twin, and automated adaptive scheduling.

Applied Materials has assembled an integrated suite of advanced manufacturing software solutions which provide a real-time, data-driven plant environment that senses factory activity, predicts performance, and prescribes effective actions. SmartFactory Rx™ improves visibility, drives quality, safety, yield and throughput, reduces cycle times and costs, and increases profitability.

Applied Materials is the leader in factory automation solutions for high volume semiconductor manufacturing plants worldwide. SmartFactory Rx, for pharmaceutical manufacturers, is powered by technologies that are customer-proven and widely deployed.

Applied SmartFactory Rx enables companies to turn their data into action:

- ***Analytics & Control***: Increase yield, prevent deviations, and assure quality
- ***Operations Productivity***: Reclaim capacity, maximize throughput, providing real-time adaptive scheduling
- ***Predictive Maintenance***: Prevent costly unscheduled downtime
- **Knowledge Advisor**: Convert operator know-how & experience into process understanding & stability

Uses cases will be presented that illustrate the kinds of problems and solutions that can be realized in both small and large molecule manufacturing plants.

580759 Machine Learning in Pharmaceutical Process Development: Where Are We up to in Pharma 4.0?

Artificial Intelligence (AI) and Machine Learning (ML) have become ubiquitous terms in the past couple of years. This presentation inspects and compares the current AI, ML approaches with alternative techniques which are already well established within the process industries, and now also in many of the innovative Pharmaceutical companies.

Advanced Process Control (APC) and Multi-Variate Analysis are data driven techniques to build models, understand the product and process, monitor for abnormal operational events and more recently directly adjust the process to achieve closed loop control of product Critical Quality Attributes (CQAs). When we compare these "traditional" data-driven techniques with Machine Learning we see the same algorithms being applied and the common goal of improved decision making through data analysis, prediction and adjustment. There is a difference however, and we'll explore what it is within the talk.

Case studies from the Pharmaceutical industry are used to demonstrate the application of several forms of Machine Learning for process control and optimisation. By comparing these Machine Learning algorithms with other tools in the process optimisation "toolbox", we can examine the benefits and challenges to implementing this technology within the industry.

581055 Digital Twins, Microfactories and People for Future Pharmaceutical Manufacturing

The pharmaceutical industry is facing increasing demands to deliver a higher variety of products more quickly, economically and sustainably. In order to achieve this, the sector is making better use out of cyber-physical production systems with a specific focus on digital twins and how they can be used to hugely minimise the amount of material utilised in the design of a manufacturing process. However, these models which will be used to build a digital twin are not focused on a single piece of equipment, process or even physics, but are holistic models which aggregate data from multiple sources. This could include for example, crystal structure, solubility predictions, population balance, CFD, and DEM. Interfacing with all this data in a manageable manner is challenging, especially for a non-modeller. This presentation highlights CMAC's recent work in 1) creating digital twins that sit across multiple model types with an interface accessible for a non-modeller, 2) undertaking end to end manufacturing of an API mirrored by an end to end process model and 3) developing and delivering training to researchers to provide them with the skills needed to design, build and operate digital twins and microfactories across pharmaceutical manufacturing.

581330 The Art of the Possible When Digitising Your Pharmaceutical Operations

The Pharma and Life Sciences sector is in the midst of rapid, fundamental change. Manufacturers are embarking on digital transformation projects across the globe but there are few that have unlocked their full potential. From Industry 4.0, Digitisation to the opportunities that Continuous Manufacturing can bring, a transformation in the operations agenda is evident. For many manufacturers, a profound transformation is required, integrating business understanding and human insights with technology innovation. We will take a look at the manufacturers who have embarked on digital transformation and are leading the way, discuss what's holding some companies back, and discuss what can be done now to stay ahead of the global competition. With a lens on emerging Industry 4.0 technologies, how Continuous Manufacturing can be an enabler of the Digitisation agenda, we will consider some of the immediate challenges and the future opportunities that will deliver 'real' business benefit and ultimately patient value.

576971 Intensification of Ammonia Production

Haber-Bosch (H-B) process is cited as one of the most important inventions of the 20th century. Ammonia is sustaining the food supply for half the population. Although carefully and completely optimized through a century of continuous effort, H-B is still very energy- and capital-intensive. High pressure requirements of Haber-Bosch process imposes substantial operating (e.g., compression) and capital (compressor cost, advanced costly alloys, thick reactor casing, etc.) expenses in the ammonia production. Cost considerations force ammonia producers to take advantage of the economy of scale to drive down the manufacture cost, while small and energy-efficient processes that can be powered with off-grid renewable energy are required for ammonia-mediated hydrogen economy. Ammonia manufacturing accounts for 1-3% of world's energy consumption, 5% of natural gas consumption, and a significant portion (~3%) of climate-changing gas emissions.3–5 These numbers incentivize us to utilize Modular Chemical Process Intensification (MCPI) methodologies to improve the energy efficiency of H-B process.

Small-scale reaction-absorption process is proposed to be a viable technology for enhanced ammonia production.[1–4] In this process, condenser is replaced by an absorber column, consisting of a packed bed of cheap, abundant metal halide salts (i.e., CaCl2) that operates at 180-250 °C. This separation unit allows lower pressure operation, which is advantageous for small-scale production. Here, we present an overview of our efforts to further intensify ammonia production via reaction-absorption process. Our target is to further improve absorbents, design catalysts that can operate at lower temperature, and optimize reaction-absorption operating conditions for lower pressure operations. Results indicate that lower pressure ($P<10$ bar) ammonia production is viable with optimizing operating absorptive separation. Better absorbents with the right chemistry and geometry, as well as optimized absorber conditions are key to achieve further improvements in production rates at lower pressure. This findings provide a more complete understanding of the proposed reaction-absorption process and proposes a path for further intensification of the ammonia production.

(1) Malmali, M.; Wei, Y.; McCormick, A.; Cussler, E. L. Ammonia Synthesis at Reduced Pressure via Reactive Separation. *Ind. Eng. Chem. Res.* **2016**, *55* (33).

(2) Reese, M.; Marquart, C.; Malmali, M.; Wagner, K.; Buchanan, E.; McCormick, A.; Cussler, E. L. Performance of a Small-Scale Haber Process. *Ind. Eng. Chem. Res.* **2016**, *55* (13).

(3) Malmali, M.; Reese, M.; McCormick, A. V.; Cussler, E. L. Converting Wind Energy to Ammonia at Lower Pressure. *ACS Sustain. Chem. Eng.* **2018**, *6* (1).

(4) Malmali, M.; Le, G.; Hendrickson, J.; Prince, J.; McCormick, A. V.; Cussler, E. L. Better Absorbents for Ammonia Separation. *ACS Sustain. Chem. Eng.* **2018**, *6* (5).

575357 Sensitivity of Desulfurization Cost for Small-Scale Natural Gas Sweetening Units

Natural gas is primarily made up of methane but also contains impurities such as sulfur.[1] The concentration of sulfur in natural gas can range from several hundred parts per million to several percent, and should be reduced below the sales gas specification limits (typically 4ppm).[1] Therefore, removing hydrogen sulfide from the natural gas, which is also called gas sweetening, is common practice. There are a variety of gas sweetening processes that are currently used by the industry such as triazine-based adsorption, Fe-Chelate Redox Systems (LO-CAT is investigated in this research), amine-based adsorption, and the Claus process. [3]

Triazine-based adsorption is generally used for removing low levels of hydrogen sulfide (<100 ppm) at relatively high gas flow rates (>20 MMSCFD) or high levels of hydrogen sulfide (>100 ppm) at relatively low gas flow rates (<20 MMSCFD).[3] The use of triazine scavengers is not recommended due to its cost when the hydrogen sulfide levels of sour gas is above 600 ppm.[3] For LO-CAT, the gas processing capacity varies from 100 to 5000 MMSCFD.[4] Amine-based adsorption is used for gas streams containing 600 ppm to 700 ppm hydrogen sulfide and limited carbon dioxide amounts.[5] Claus sulfur recovery plants can decrease the hydrogen sulfide levels to 250 ppm with gas processing capacities greater than 4500 MMSCFD.[6] Additional sulfur removal processes may be required after amine-based adsorption and Claus sulfur recovery plants because hydrogen sulfide levels of the treated gas may be higher than tail gas specifications at large gas flow rates.[6] A new small-scale desulfurization process, SourCat,[7] is under-development by Sour Gas LLC and IntraMicron, Inc., an Auburn University spin-off. The feasibility of natural gas sweetening from distributed small-scale production wells, i.e., from stranded resources, depends on the desulfurization costs of these technologies. Among the desulfurization processes discussed, triazine-based adsorption, LO-CAT and SourCat can be used for small-scale gas sweetening operations.

In this work, a systematic sensitivity analysis is performed to investigate how desulfurization costs of the triazine-based adsorption process, LO-CAT, and SourCat change with changes in inlet sour gas specifications, utility prices and raw material costs at small scales relevant for stranded gas sweeting operations. Sour gas specifications considered are gas flow rate (100 MSCFD – 100 MMSCFD) and hydrogen sulfide concentration (500 ppm – 2000 ppm). Process simulations of these three gas sweetening processes are built using Aspen Plus. Capital and operating costs are calculated using an economic analysis tool called ECON.[8] Desulfurization costs ($/MSCF natural gas) are then calculated before and after the payback period (which is assumed to be three years). This presentation will discuss in detail the robustness of the desulfurization costs for these three processes with respect to the changes in the inlet conditions, raw material costs and utility prices.

References

1. Madsen, H. T. "Investigation of fouling formation during H2S scavenging with 1, 3, 5-tri-(2-hydroxyetthyl)-hexahydro-s-triazine." *Esbjerg Institute of Technology, Aalborg University* (2011).
2. Wanek, Rick. "Monitoring H2S to meet new exposure standards." Occupational health & safety (Waco, Tex.) 80.9 (2011): 20-22.
3. Fundamentals of Low-Tonnage Sulfur Removal and Recovery. Echt, B., D. Leppin, D. Mamrosh, D. Miradian, D. Seeger, and B. Warren.February 26, 2017. "Fundamentals of Low-Tonnage Sulfur Removal and Recovery." Presented at the Laurance Reid Gas Conditioning Conference, Norman, OK.
4. Rouleau, W., – Small Capacity Sulfur Recovery Units for Coal Gasification, Merichem Company, 2011
5. Ankur J., Cameron. Wanted: Efficient Gas Treatment System To Fit Tight FPSO Footprint, HARTENERGY Company, 2015
6. Cheremisinoff, Nicholas P., and Paul E. Rosenfeld. Handbook of Pollution Prevention and Cleaner

Production Vol. 1: Best Practices in the Petroleum Industry
7. Catalyzing Commercialization: Sour Gas Has a Sweeter Future. (2017, January). Retrieved from https://www.aiche.org/resources/publications/cep/2017/january/catalyzing-commercialization-sour-gas-has-sweeter-future
8. Kalakul, S., Malakul, P., Siemanond, K. and Gani, R., 2014. Integration of life cycle assessment software with tools for economic and sustainability analyses and process simulation for sustainable process design. Journal of cleaner production, 71, pp.98-109.

565616 Dynamic Intensification of Ternary Distillation Columns

Process intensification (PI) is a set of innovative design principles that transcend the traditional unit operations framework. One of the key tenets of PI is bringing multiple phenomena into closer physical proximity [1], thereby achieving smaller, safer and more efficient processes. To date, the vast majority of process intensification strategies consist of design changes, focusing on novel devices and constructs that implement the aforementioned tenet in practice. Such devices have clear advantages, in terms of, e.g., reduced transport limitations. However, tight dynamic coupling between different phenomena can also lead to control challenges, such as a reduction in the number of degrees of freedom available for control, as well as a narrower operation window compared to conventional processes [2]. Furthermore, the implementation of such new design also implies major capital expenditures.

In this paper, we will rely on a strategy for process intensification in the temporal domain, which we refer to as "dynamic process intensification (DPI)." DPI seeks to achieve improvements in process efficiency and economics by exploiting and favorably manipulating the interplay of multiple coupled phenomena occurring in a conventional or intensified process [3]. In particular, DPI relies on periodic forcing on the available manipulated variables of a column, imposing a cyclic operating regime that generates the desired product as a mixture of two or more auxiliary products, each having lower energy consumption than the desired product [4,5].

Our research is motivated by the large energy consumption of multi-component (in particular, ternary) distillation columns across the chemical and petrochemical industries. We present a conceptual framework for dynamic intensification. Then, the canonical model of a ternary distillation column separating a mixture of cyclohexane, toluene and m-xylene is used to illustrate the theoretical ideas. We demonstrate that, all other factors being equal, dynamic process intensification results in considerable energy savings compared to steady-state operation of the same column, with the added economic benefit that it requires minimal capital expenditure.

References

[1] Creative Energy. European Roadmap for Process Intensification, Technical report, Creative Energy, 2007.

[2] Schembecker, G.; Tlatlik, S. Process Synthesis for Reactive Separations, Chem. Eng. Process, 42, 179-189, 2003.

[3] Baldea, M.; Edgar, T. F. Dynamic Process Intensification, Dynamic Process Intensification, Current Opinion in Chemical Engineering, 22, 48-53, 2018

[4] Yan, L.; Edgar, T. F.; Baldea, M. Dynamic Process Intensification of Binary Distillation Based on Output Multiplicity, AIChE J, 65 (4), 1162-1172, 2019.

[5] Yan, L.; Edgar, T. F.; Baldea, M. Dynamic Process Intensification of Binary Distillation via Periodic Operation, Ind. Eng. Chem. Res., 2018. https://doi.org/10.1021/acs.iecr.8b04852.

573731 Towards Safe and Operable Process Intensification Systems

The challenges that process industries face to enhance their economic, environmental and safety performance have led to the development and incorporation of Process Intensification (PI) solutions. Several Process Systems Engineering (PSE) approaches have been proposed in the literature to address the synthesis, design, operation and optimization of PI systems [1,2,3,4,5]. As far as the operability of these systems is concerned, the combination of multiple process operations in a single unit results in a reduction of the degrees of freedom in comparison to having consecutive process units, which limits control opportunities and rises safety concerns [6,7]. Apart from that, the steady state and dynamic models that are used for the description of intensified systems are coupled and highly nonlinear, making the construction of accurate approximation representations to reduce the computational complexity, challenging [8]. Thus, the derivation of operable PI designs which incorporate safety considerations is critical and still an active area of research.

This work presents a strategy for the systematic assessment of the safety and operability performance of intensified process systems. Utilizing the PARametric Optimization & Control (PAROC) Framework [9], we are able to benchmark and compare PI systems. Firstly, we develop high-fidelity dynamic models of conventional and the equivalently derived intensified designs. Based on these models, we derive optimal control policies using multiparametric programming and define and compare the regions of operation of the aforementioned systems. Safety considerations are incorporated in the control design problem from which the failure and consequence severity are quantified [7,10]. The proposed approach is highlighted through a reactive separator case study.

References:

[1] Tian, Y., Demirel, S. E., Hasan, M. F., & Pistikopoulos, E. N. An overview of process systems engineering approaches for process intensification: state of the art. Chemical Engineering and Processing-Process Intensification, 2018. **DOI:** 10.1016/j.cep.2018.07.014.

[2] Papalexandri, K. P., & Pistikopoulos, E. N. Generalized modular representation framework for process synthesis. AIChE Journal, 1996, 42(4), 1010-1032.

[3] Demirel, S. E., Li, J., & Hasan, M. F. Systematic process intensification using building blocks. Computers & Chemical Engineering, 2017, 105, 2-38.

[4] Pichardo, P., & Manousiouthakis, V. I. Infinite DimEnsionAl State-space as a systematic process intensification tool: Energetic intensification of hydrogen production. Chemical Engineering Research and Design, 2017, 120, 372-395.

[5] Tula, A. K., Babi, D. K., Bottlaender, J., Eden, M. R., & Gani, R. A computer-aided software-tool for sustainable process synthesis-intensification. Computers & Chemical Engineering, 2017, 105, 74-95.

[6] Baldea, M. From process integration to process intensification. Computers & Chemical Engineering, 2015, 81, 104-114.

[7] Tian, Y., & Pistikopoulos, E. N. Synthesis of Operable Process Intensification Systems—Steady-State Design with Safety and Operability Considerations. Industrial & Engineering Chemistry Research, 2019. **DOI:** 10.1021/acs.iecr.8b04389.

[8] Nikacevic, N. M., Huesman, A. E., Van den Hof, P. M., & Stankiewicz, A. I. Opportunities and challenges for process control in process intensification. Chemical Engineering and Processing: Process Intensification, 2012, 52, 1-15.

[9] Pistikopoulos, E. N., Diangelakis, N. A., Oberdieck, R., Papathanasiou, M. M., Nascu, I., & Sun, M. PAROC—An integrated framework and software platform for the optimisation and advanced model-based control of process systems. Chemical Engineering Science, 2015, 136, 115-138.

[10] Freeman, R. A. (1990). CCPS guidelines for chemical process quantitative risk analysis. Plant/Operations Progress, 9(4), 231-235.

560846 Reprocessable 3D-Printed Conductive Elastomeric Composite Foams for Strain and Gas Sensing

Mechanically robust, highly-compressible, and conductive thermoplastic polyurethane/carbon black foams were successfully fabricated via 3D printing. Thixotropic inks were formulated by dispersing nanoclay, carbon black, and polymer in an organic solvent and then 3D multifunctional sensors were prepared by direct ink writing (DIW). Sequential removal of solvent and nanoclay yielded a porous polymer/carbon black structure which maintained excellent elasticity and compression cyclability. The presence of carbon black in the foams led to good electrical conductivity and stable piezoresistive sensing signals at a strain of up to 80%, including the ability to distinguish human motions. Complementary to a decrease in resistance upon compression, the foam can be used as a gas sensor device, as exposure to volatile organic compounds causes the polymer to swell and an increase in resistance. The printed conductive foams can be easily recycled and reprocessed by dissolution in an organic solvent and subsequent re-3D printing, with little detriment to performance.

570244 Computational Printing of Cement-Based Pastes with Time-Dependent Rheology

Recent advances in additive manufacturing (AM or 3D printing) of cement-based structures has led to new rheological and set (hydration) requirements for fresh cementitious pastes. 3D printing of cement-based materials requires four primary paste-related processing characteristics: (1) extrusion, (2) layer-by-layer deposition, (3) resistance to deformation under layering loads, and (4) time response of the printing paste, i.e. hydration rate and time-dependent rheological properties. Designing effective cement-based printing pastes involves optimization of paste yield stress, plastic viscosity and thixotropic response and design strategies and data correlating rheological properties to printability are among the existing research gaps. Thus, the objective of this study is to link relevant experimentally determined hydration kinetics and time-dependent paste rheological parameters, including yield stress, viscosity and structuration, to computational printing frameworks in an effort to correlate properties to printability metrics and to target optimal paste design parameters. Prior to conducting simulations, rheometric and calorimetric studies were done to define and extract rheological fluid properties as a function of extent of hydration. Computational fluid dynamics (CFD) simulations were then developed for printing of fresh cement paste in 2D and 3D geometries. Free-surface flow of square, spherical, and rectangular gobs of materials in 2-D geometries were simulated as a function of fluid properties linked to hydration kinetics. The importance of rheological characteristics and rate of hydrations (phase change) were quantified by deformation and rate of deformation of simulated cement paste gobs in both 3D and 2D geometries respectively. Cylinders and cone benchmark objects were also printed for validation.

576584 Selective Nanoparticle Orientation for Polymer Composites Via Hybrid Manufacturing

Nanoparticle alignment can benefit the controllable light transmittance, targeted drug delivery, gaseous or liquid separation/purification, localized heat transfer, and, the mechanical reinforcement in composites. Self-assembly, electrical or magnetic fields, and mechanical forces have been used to achieve the nanoparticle arrangements. However, it is hard to reach large manufacturing speeds and high fabrication resolutions. Our approach using 3d printing and template effects allows producing scalable preparation of samples on which nanoparticle orientations can be well controlled. The in-house designed 3d printing had the developed surface topologies with a series of patterns such as line grooves or dot wells manufactured on substrates. Both resin-based and filament-based additive manufacturing mechanisms were included in the manufacturing process. Upon dispersed in suitable solvents, the nanoparticles adhered to the printed surface along the liquid-air-solid contact lines. The drying thermodynamics and the confinement effects from the template arranged the nanoparticles with different morphologies. Additive layers were duplicated on top of each polymer/nanoparticle pair, and results of the nanoparticle alignment were examined. Both thermoplastic and thermoset polymers were used as substrates, and surface tension influences on nanoparticle arrangements were also correlated. The orientation of nanoparticles and the relationships to the processing conditions were studied by polarized optical microscopy (POM), atomic force microscopy (AFM), x-ray diffraction (XRD), scanning electron microscopy (SEM) and transmission electron microscopy (TEM). Mechanical properties like static modulus, strength and toughness were measured for the layered composites.

577397 Two-Color Photoinitiation / Photoinhibition for Continuous Stereolithographic Additive Manufacturing

Continuous stereolithographic additive manufacturing allows high print speed and eliminate directionally-dependent material properties present in traditional layer-by-layer approaches. Current approaches allow continuous printing by eliminating adhesion to the print window during printing (which necessitates separation and recoating steps) by inhibiting polymerization in a thin region, or "dead zone", using oxygen inhibition.

Another method to enable continuous 3D printing is to use a two-color polymerization system where one wavelength initiates polymerization and a secondary orthogonal wavelength inhibits polymerization. This approach enables increased control over the dead zone and allows additional functionality over what is possible with the conventional oxygen-inhibited.

The theoretical and practical aspects of this approach are presented along with the promise and challenges of this novel continuous additive manufacturing approach.

578781 Direct Ink Writing of Modified Cellulose Nanocrystals

Cellulose nanocrystals (CNCs) are stiff, lightweight rod-shaped nanoscale materials that have a tensile strength eight times that of steel (7.5 GPa). In addition to their outstanding mechanical properties, CNCs are low cost, low density (~1.6 g/cm^3), renewable in nature, and are easily processed. These nanoparticles have attracted the attention of researchers and there has been an exponentially increasing number of works devoted to understanding such materials and their applications. The addition of CNCs into a polymer matrix has been shown to enhance the mechanical properties of the resulting composite, however, the utilization of CNCs in more common non-polar polymers, such as epoxy resins, is still limited to their hydrophilicity and poor dispersibility. Surface modification is hence needed to aid in dispersion. With the proper modification, dispersion of these CNCs into common non-polar polymers, such as epoxy resins, polypropylene, or polystyrene, can allow for the development of robust, durable, lightweight, and transparent materials to be used in structural composites, adhesives, energy storage devices, actuators, and sensors. Understanding the dispersibility, structure of CNC's in the matrix and its resultant effect on macroscopic properties such as rheology and hence printing performance is of utmost importance before these composites are put to use.

In this work we characterize the effects of a functionalized CNC on the rheological properties and printing behavior (Direct Ink Writing (DIW)) of an epoxy resin. CNC functionalization is achieved by grafting an activated medium chain fatty acid to substitute the hydroxyl functional groups. The level of CNC modification is evaluated using Fourier Transform Infrared (FTIR), thermo-gravimetric analysis (TGA), X-ray diffraction (XRD), and solvent dispersibility studies. Functionalized CNCs are dispersed into the epoxy resin (EPON 828) at loadings up to 20 wt%. Initially, rheological studies are used in this study to assess the printability of the inks. The rheological percolation threshold is the critical concentration at which there is a transition from liquid-like to solid-like behavior. For this system, this value was found to be at ~10wt%. We found that for the final printed structure to exhibit structural integrity, it is necessary to print at loadings slightly above this value. Future work will include microscopy (SEM and polarized light) of both printed and cast samples to observe the dispersion and alignment of the CNCs within the polymer matrix. Finally, the mechanical properties of all samples will be measured to show the effect of the reinforcement in the matrix.

579268 A Review and Outlook on Machine Learning and Data Science in Chemical Engineering

With today's development of the internet, internet of things, wireless sensors and communications, smart manufacturing, and high-throughput experiments, the amount of data collected has grown exponentially, analogous to the Moore's law in the increase of solid-state transistor density. The data explosion has made all sectors including engineering, medicine, business, commerce, finance, and materials science to endorse the potential and power of big data. The relevance of big data to chemical engineering and science is evident from the recent technical contributions by chemical engineers and scientists.

Great successes in machine learning today have to do with natural language processing (NLP), multi-lingual translation, image recognition, etc. In addition, the demonstration of AlphaGo by Google's DeepMind in early 2016 was one of the most shocking events to the world. However, the game of Go has little in common with the degree of uncertainty and the complexity of real-world engineering problems. In this talk we discuss the characteristics of process systems engineering problems and provide a brief review of machine learning and data analytics in chemical engineering in the last three decades. We give a few attributes for machine learning techniques to be widely applicable in the process systems. Owing to these characteristics, process monitoring and inferential sensors are two of the most adopted data-driven applications in chemical engineering.

The talk will then provide an assessment of the resurgent interest in deep neural networks and deep reinforcement learning and contrast them against the technology two decades ago. We also give a few examples of high-profile applications that failed the test of time. We assert that most of the sustainable applications are domain-specific, in that they incorporate specific structures and knowledge in the models. We will briefly touch on the tremendous advancement in statistical machine learning, which are the core of data science and could propel the advances in other scientific and engineering disciplines, such as smart manufacturing, bioinformatics and materials science.

559766 Neural Network-Based Modeling and Operation for ALD of SiO_2 Thin Films Using Data from a Multiscale CFD Simulation

Atomic layer deposition (ALD) plays an important role in the manufacturing of highly conformal thin-film materials, which are used extensively in the semiconductor industry. In our previous study, a multiscale computational fluid dynamic (CFD) model has been developed for the SiO2 ALD process in an optimized showerhead ALD reactor geometry. The optimized reactor design involves a volume adjusting horn in the upstream region and a radially adaptive sizing of the showerhead holes [1], [2] . The results from the developed macroscopic CFD model show that those two design features enable quick achievement of steady-state in the gas-phase domain of the reactor, and provide a relatively uniform surface pressure and temperature profile for the microscopic kinetic Monte-Carlo (kMC) modeling of the surface deposition, which authentically simulates the film deposition process [3].

However, the multiscale CFD model is computationally demanding and is not suitable for applying real-time control schemes like model-based controllers (MPC). Motivated by this, in this work, we investigate the transient operation of the full reactor in details, using the model developed previously. The impact of operation parameters, including inlet process gas flowrate, temperature and pressure, on the deposition process is discussed and a corresponding database with temporal information of the process is constructed. Also, the appropriate operating boundary accounting for both microscopic and macroscopic operations is determined to ensure process feasibility [3]. Subsequently, a recurrent neural network (RNN) is constructed and trained using the model-generated database, which enables the prediction on the transient process dynamics based on the inputs of operating condition [4]. Moreover, utilizing the RNN model, a control scheme is designed based on the developed RNN to reduce the impact of potential disturbances on the deposition process and enhance the film uniformity. Finally, an operation example is shown to demonstrate the practicability of the proposed method.

[1] Lee, C.S., Oh, M.S., Park, H.S. US Patent No. US7138336B2. 2003. Retrieved from

https://patentimages.storage.googleapis.com/01/e8/3e/5c751ffe023e4d/US7138336.pdf

[2] Crose, M., Zhang, W., Tran, A., Christofides, P.D. Multiscale three-dimensional CFD modeling for PECVD of amorphous silicon thin films. Computers & Chemical Engineering. 2018, 113, 184 – 195.

[3] Ding, Y., Zhang, Y., Kim, K., Tran, A., Wu, Z., Christofides, P.D. Microscopic Modeling and Optimal Operation of Thermal Atomic Layer Deposition. Chemical Engineering Research and Design. 145, 159-172, 2019.

[4] Connor, J.T., Martin, R.D., Atlas, L.E. Recurrent neural networks and robust time series prediction. IEEE Transactions on Neural Networks. 1994, 5, 240-254.

579520 Keynote Talk: Intelligent Cloud-Based Algorithms for Pharmaceutical Drug Development

The ability to predict or explore polymorphism in organic crystal compounds is of great importance for many industries since many physical and chemical properties of molecular crystals strongly depend on the molecular packing in the solid state. Crystal structure prediction (CSP) aims to reduce the risk and improve the efficiency of pharmaceutical drug manufacturing by correctly predicting the low energy crystal forms of organic molecules from minimal information, such as the chemical diagram. At XtalPi, we have developed a cloud-based computational platform that combines advanced physics-based algorithms with A.I/machine learning algorithms. Our CSP platform can rapidly generate and rank millions of crystal structures for pharmaceutically relevant systems and we highlight some cases in which our CSP platform has revealed low energy polymorphs in challenging systems. We will also discuss some areas where we are expanding our computational platform to make accurate predictions on the physiochemical and pharmaceutical properties of small-molecule candidates for drug design, solid-form selection, and for other critical aspects of drug development.

568479 Industrial Success Stories with Machine Learning and Deep Learning in Matlab

Machine learning has rapidly gained in popularity, spurred by well-publicized advances like deep learning and widespread commercial interest in big data analytics. Despite the enthusiasm, companies and researchers are struggling to deploy machine learning models in chemical plants so as to reconcile the hype with actual application and success.

In this presentation, MathWorks will showcase the successes of machine learning and deep learning in industrial settings with 6 user stories/case studies - all drawn from the process industries. With these examples drawn from customers' usage of MATLAB, MathWorks seeks to dispel the notion that machine learning is solely the subject of academic study but can indeed be applied to process plants with measurable business benefits.

In addition, machine learning workflow elements will be discussed in this talk including pre-processing, model building using apps and deployment to industrial systems using MATLAB.

569189 Keynote Talk: Modern Data Analytics to Accelerate the Design of Advanced Materials

We will discuss three critical aspects of applying a machine learning approach to accelerate the design of advanced materials for energy applications: 1) consistently measured, relevant experimental data with known pedigree, 2) in-depth scientific features that can be correlated with target properties, and 3) training of surrogate machine learning models. We will present examples of predicting mechanical properties and oxidation response of high-temperature alloys, thermochemistry of complex oxides, and melt-pool microstructure of single-track by the additive manufacturing process.

The research was sponsored by the LDRD Program of Oak Ridge National Laboratory and the Department of Energy, Vehicle Technologies Office, Propulsion Materials Program.

570365 GPU-Accelerated Computation of Stirred Tanks

A fast and economical solver, accelerated by the Graphics Process Units (GPU) of a single graphics card in a desktop computer, is developed for the simulation of stirred tanks, integrating Lattice Boltzmann Method (LBM), coherent Large Eddy Simulation (LES) and Immersed Boundary Method (IBM). The grid resolution can reach 13.8 million nodes in maximum, resolving considerable details of flow field in stirred tanks with only such a simple desktop computer. In the meantime, the computational speed is 1500-fold faster than that of the traditional transient simulation of Computational Fluid Dynamics (CFD) based on Navier-Stokes equations implemented on 16 CPU cores. It only takes less than two minutes to update the simulation of one impeller revolution, enabling the transient simulation of longer physical time of stirred tanks. We find that at least 4,000 impeller revolutions in simulation is needed to achieve an obviously time-independent macro-instability frequency. With the coherent LES model and reasonable grid resolution, the fast solver is able to resolve more than 95% of total Turbulent Kinetic Energy (TKE) for highly turbulent region, reproducing the monotonic decrease of TKE along the radial direction, much better than the Smagorinsky model. The average velocity and TKE predicted by the fast solver are in better agreement with the experimental data in literature. The solver is therefore more suitable for the fast simulation of stirred tanks using only a desktop computer without the need of much finer grid resolution and supercomputers. We believe this high-resolution approach serves as a good gateway to further integrate with Artificial Intelligence for chemical reactors.

564376 Keynote Talk: Faster Prediction with Artificial Intelligence: Using Process Simulations and Production Data to Develop Fast-Running Inference Models for Manufacturing

Navigating a large-dimensional design space with simulations is a significant improvement over doing so with physical experiments but can still be very time consuming. Deep learning methods are often used to develop fast-running surrogate models of more computationally expensive simulations in order to get near-real-time predictions and to optimize complex systems such as manufacturing processes, energy systems, and fusion reactors. In this talk, I will discuss an application of machine learning to develop a fast-running surrogate models that captures the dynamics of industrial multiphase fluid flows. I will also discuss an improved population search method that can help the analyst explore a high-dimensional parameter space to optimize production while reducing the model uncertainty. Results from demonstration problems and from real-world Inertial Confinement Fusion simulations will be presented.

569806 Understanding the Role of Partial Oxidation Reactions during Autothermal Pyrolysis

The goal of this project is to understand the partial oxidation reactions responsible for energy release during biomass autothermal pyrolysis. The enthalpy for pyrolysis is provided by exothermic oxidation reactions within the reactor instead of external energy supplied by heat transfer. This is achieved by reacting a small amount of oxygen (equivalence ratios of 0.06-0.10) with biomass in a fluidized bed reactor, which partially combusts the products of pyrolysis to overcome the endothermic enthalpy of pyrolysis. Experiments performed at Iowa State have shown that autothermal pyrolysis does not produce significant loss in bio-oil yield or quality as compared to traditional non-oxidative pyrolysis. Consequently, we are developing a chemical kinetic model of autothermal pyrolysis to better understand the source of energy for the process.

Based on product differences between autothermal pyrolysis and inert pyrolysis (bio-oil, biochar, and gases), it appears the enthalpy for pyrolysis is being provided primarily by the oxidation of char, light oxygenates, and phenolic compounds. However, to confirm these products are in fact oxidizing at pyrolysis temperatures (400-600 °C) further experiments are needed. To accomplish this, model compounds representative of pyrolysis products such as acetic acid, levoglucosan, and phenolic monomers are partially oxidized at pyrolysis temperatures in a modified isothermal fluidized bed reactor. The condensed liquid product and non-condensable gases are analyzed via gas chromatography (GC) to determine products of partial oxidation and estimate kinetic data. Preliminary results show that these products begin to oxidize at temperatures below 400 °C, much lower than current published reaction mechanisms predict. Likewise, the role of incomplete combustion that does not generate CO_2 and H_2O appears to occur at these low temperatures and equivalence ratios. In particular, the lignin-derived phenolic components appear to incorporate oxygen into their structures during partial oxidation rather than generate non-condensable gases. Char oxidation studies were also performed using both thermogravimetric analysis (TGA) and fluidized bed experiments. Both methods indicate that the amount of alkali and alkaline earth metals (AAEMs) present in the biochar influence rates of reaction and product composition (CO vs CO_2). Determining the relationship between product oxidation at pyrolysis temperatures can identify not only the source of energy in biomass autothermal pyrolysis, but potentially expands autothermal processing to other applications as well.

576533 Structural and Chemical Changes in Plant Cell Walls during Early Stages of Thermal Deconstruction

Volatile products from lignocellulosic biomass thermal deconstruction processes—combustion, gasification, fast pyrolysis, and solvent liquefaction—have been well characterized, but the solid and liquid phase reactions that occur in the early stages of decomposition are largely unknown. Here we analyze the initial solid phase biomass thermal deconstruction reactions *in planta* and with high particle heating rates, creating the most accurate investigation to date of how these processes occur. Using a variety of instrumentation, we quantify the extent and relative rates of deconstruction, demonstrating that biopolymers resist the thermally energetic conditions to differing degrees, even when ensconced in biomass cell walls. Hemicellulose and the more frangible lignin components decompose and volatilize more readily than cellulose; this outcome temporarily enriches biomass with cellulose. These chemical changes additionally manifest in larger cell wall structure and mechanical property transformations. In all, our investigation concludes that these solid phase reactions strongly influence the production rates of volatile species and will require additional study before these processes can be modeled precisely or desired product yields potentially improved.

579943 Multiscale Simulation of Autothermal Pyrolysis of Biomass

The variability of biomass substantially exceeds that of conventional feedstocks such as coal due to inherent biocomplexities and inter-species diversity. The impacts of feedstock attributes which are unique to biomass, such as anisotropic, species-specific microstructure, highly variable ash and moisture content, and relatively large particle size distributions are well- recognized by the biomass conversion community. Problems associated with the complexity of biomass feedstocks have been acknowledged by the combustion and biopower communities. However, conventional simulation frameworks for thermochemical conversion of biomass feedstocks employ overly-simplified models that largely neglect the aforementioned complexities and/or apply generalized heuristics which limit their utility to the single feedstock and reactor system for which they were developed. In this talk will describe the development of a multiscale simulation approach that account for the complex size, shape, and chemistry of various biomass feedstock sources. The modeling framework is based on our previously-developed biomass particle pyrolysis simulations which account for the complex, anisotropic transport phenomena exhibited by biomass during thermochemical conversion processes. These particle scale models are coupled to reactor-scale simulations to account for bulk hydrodynamics and residence time distributions. In addition to these physics, we have implemented of a kinetic scheme suitable for biomass combustion and autothermal pyrolysis. I will also provide an update on our efforts to enable direct import of 3D image data of biomass particles into simulation environments and the implementation of adaptive mesh refinement to capture geometric changes that result from pyrolysis and combustion.

576446 Significant Increase in the Production of Sugars from Pyrolysis of Agricultural Residues By Application of Iron Sulfate Pretreatment

Fast pyrolysis of lignocellulosic biomass produces very little sugar or anhydrosugars compared to pyrolysis of pure polysaccharides because naturally occurring alkali and alkaline earth metals (AAEMs) in biomass catalyze the fragmentation of pyranose and furanose rings. Sugar yields can be dramatically increased by pretreating the biomass with sulfuric acid prior to pyrolysis, which passivates the catalytic activity of the metals by converting them into thermally stable salts. However, these AAEMs are also important to the depolymerization of lignin. Their passivation has the unintended consequence of slowing the rate of volatiles release from lignin, which results in a transient melt phase that agglomerates and fouls fluidized bed pyrolyzers. To overcome this problem, an alternative catalyst that is active toward lignin depolymerization but not pyranose ring fragmentation is needed. Pretreatment of herbaceous biomass with ferrous sulfate was found to have the desired effects of both passivating AAEMs and introducing iron that is catalytic toward lignin depolymerization. Conventional nitrogen-blown pyrolysis of pretreated corn stover increased the reactor weight hourly space velocity (WHSV), defined as biomass feed rate (kg/hr) per fluidized bed mass (kg), from 1 hr^{-1} with acid treatment to 4 hr^{-1} with iron sulfate. Substituting ferrous sulfate for sulfuric acid as pretreatment of corn stover had an even more dramatic increase in WHSV when operating the pyrolyzer under air-blown (autothermal) conditions, increasing from 3 $hr^{-1}$to 13 hr^{-1}. Sugar yields from the pyrolysis of corn stover increased from 0.5 wt% to 13 wt% on a biomass basis – a 26-fold increase through the use of the ferrous sulfate pretreatment. These advantages were combined to increase volumetric sugar productivity by two orders of magnitude, from 30 g/L/hr for conventional untreated corn stover to 3000 g/L/hr for autothermal ferrous sulfate treated corn stover.

579126 A Novel Fluidized Bed Model for Fast Pyrolysis Applications

Biomass fast pyrolysis has been identified as a viable method to produce bio-oil, a renewable alternative to petroleum-derived fuels and feedstocks. Fluidized bed reactors are ideal for pyrolysis due to their precise temperature control and high solid-solid heat transfer rates, and recent developments in autothermal operation have overcome the limitations inherent in supplying externally generated heat to the reactor itself. While many fluidized bed reactor models have been developed that accurately depict fluid-phase or heterogeneous reactions, they do not capture the intricacies of solid-phase reactions where heat transfer is the limiting factor.

This work presents a novel approach to the modeling of a fluidized bed reactor, focusing on the experience of the particles being pyrolyzed, which make up a minority of the bed mass. Due to the decoupling of solid- and vapor-phase reactions, this subset of particles can be modeled as a series of open control volumes surrounded by a uniform temperature heat source. The phenomena of heat transfer, chemical reaction, density loss, and elutriation are all captured in a format applicable within commercially available process simulators. The addition of partial oxidation reactions allows the reactor model to operate autothermally, with results supporting experimental findings on the performance enhancement derived from autothermal operation.

579942 CFD Investigation of Autothermal Biomass Pyrolizers

Autothermal biomass pyrolysis processes [1] are being developed by reaction engineers to address one of the main bottlenecks to the scale-up of biomass pyrolysis, which is the heat transfer limitation [1,2] due to the small heat transfer flux that can be sustained in the fluidized bed pyrolyzers used to conduct the process. Autothermal pyrolysis aims at addressing this limitation by allowing partial oxidation [1,3] of pyrolysis products, which allows the in-situ generation of heat needed to sustain the pyrolysis process.

The performance of autothermal pyrolizers may be affected by several design choices and by the composition of the biomass feed stream. The geometric configuration of the pyrolyzer and the location of the biomass feed may affect mixing of the biomass in the pyrolizer, impacting the residence time of the biomass and the product yield.

In order to investigate these effects, we have formulated a computational model to describe biomass fast pyrolysis. The model is based on the Euler-Euler multi-fluid model [4,5] with kinetic theory closures for polydisperse granular phases [6,7], supplemented by a comprehensive chemical kinetic mechanism [8–18] accounting for devolatilization, char combustion and gas-phase reactions. Results demonstrating the predictive capabilities of the computational models in comparison to experimental data of pyrolysis product yield will be presented.

The model was then used to study mixing of biomass and sand in a fluidized bed pyrolizer to understand the level of mixing obtained in the fluidized bed with different injection location

of the biomass. To this purpose, a mixing index insensitive to the spatial discretization of the computational domain was formulated, in order to systematically quantify the quality of the mixing of the granular mixture. Numerical experiments were first performed considering a cold flow in a laboratory scale fluidized bed with sand and biomass without accounting for the effect of eventual effects of the temperature distribution and of the chemical reactions.

The effects of heat transfer and chemical reaction are then introduced to investigate the impact of the location of the injection point of the biomass on the products in realistic operating conditions.

References

[1] J.P. Polin, C.A. Peterson, L.E. Whitmer, R.G. Smith, R.C. Brown, Process intensification of biomass fast pyrolysis through autothermal operation of a fluidized bed reactor, Appl. Energy. 249 (2019) 276–285. doi:10.1016/j.apenergy.2019.04.154.

[2] J. Proano-Aviles, J.K. Lindstrom, P.A. Johnston, R.C. Brown, Heat and Mass Transfer Effects in a Furnace-Based Micropyrolyzer, Energy Technol. 5 (2017) 189–195. doi:10.1002/ente.201600279.

[3] K.H. Kim, X. Bai, M. Rover, R.C. Brown, The effect of low-concentration oxygen in sweep gas during pyrolysis of red oak using a fluidized bed reactor, Fuel. 124 (2014) 49–56. doi:10.1016/j.fuel.2014.01.086.

[4] D.A. Drew, Averaged equations for two-phase flows, Stud. Appl. Math. L (1971) 205 – 231.

[5] D.A. Drew, Continuum Modeling of Two-Phase Flows, in: R. Meyer (Ed.), Theory Dispersed Multiph. Flow, Academic Press, 1983: pp. 173 – 190.

[6] J.T. Jenkins, S.B. Savage, A theory of the rapid flow of identical, smooth, nearly elastic spherical particles, J. Fluid Mech. (1983) 187 – 202.

[7] J.T. Jenkins, F. Mancini, Balance Laws and Constitutive Relations for Plane Flows of a Dense, Binary Mixture of Smooth, Nearly Elastic, Circular Disks, J. Appl. Mech. 54 (1987) 27–34. doi:10.1115/1.3172990.

[8] M. Calonaci, R. Grana, E. Barker Hemings, G. Bozzano, M. Dente, E. Ranzi, Comprehensive Kinetic Modeling Study of Bio-oil Formation from Fast Pyrolysis of Biomass, Energy Fuels. 24 (2010) 5727–5734. doi:10.1021/ef1008902.

[9] M. Corbetta, A. Frassoldati, H. Bennadji, K. Smith, M.J. Serapiglia, G. Gauthier, T. Melkior, E. Ranzi, E.M. Fisher, Pyrolysis of Centimeter-Scale Woody Biomass Particles: Kinetic Modeling and Experimental Validation, Energy Fuels. 28 (2014) 3884–3898. doi:10.1021/ef500525v.

[10] E. Ranzi, P.E.A. Debiagi, A. Frassoldati, Mathematical Modeling of Fast Biomass Pyrolysis and Bio-Oil Formation. Note I: Kinetic Mechanism of Biomass Pyrolysis, ACS Sustain. Chem. Eng. 5 (2017) 2867–2881. doi:10.1021/acssuschemeng.6b03096.

[11] R.S. Miller, J. Bellan, A generalized biomass pyrolysis model based on superimposed cellulose, hemicellulose and lignin kinetics, Combust. Sci. Technol. 126 (1997) 97–137. doi:10.1080/00102209708935670.

[12] C. Di Blasi, Heat, momentum and mass transport through a shrinking biomass particle exposed to thermal radiation, Chem. Eng. Sci. 51 (1996) 1121–1132. doi:10.1016/S0009-2509(96)80011-X.

[13] R. Radmanesh, Y. Courbariaux, J. Chaouki, C. Guy, A unified lumped approach in kinetic modeling of biomass pyrolysis, Fuel. 85 (2006) 1211–1220. doi:10.1016/j.fuel.2005.11.021.

[14] M.R. Rover, P.A. Johnston, L.E. Whitmer, R.G. Smith, R.C. Brown, The effect of pyrolysis temperature on recovery of bio-oil as distinctive stage fractions, J. Anal. Appl. Pyrolysis. 105 (2014) 262–268. doi:10.1016/j.jaap.2013.11.012.

[15] A. Williams, M. Pourkashanian, J.M. Jones, Combustion of pulverised coal and biomass, Prog. Energy Combust. Sci. 27 (2001) 587–610. doi:10.1016/S0360-1285(01)00004-1.

[16] R. Barranco, A. Rojas, J. Barraza, E. Lester, A new char combustion kinetic model 1. Formulation, Fuel. 88 (2009) 2335–2339. doi:10.1016/j.fuel.2009.02.005.

[17] C. Di Blasi, Combustion and gasification rates of lignocellulosic chars, Prog. Energy Combust. Sci. 35 (2009) 121–140. doi:10.1016/j.pecs.2008.08.001.

[18] G. Perkins, Underground coal gasification – Part II: Fundamental phenomena and modeling, Prog. Energy Combust. Sci. 67 (2018) 234–274. doi:10.1016/j.pecs.2018.03.002.

575056 Developing a Systematic Approach for the Collaboration between Process Systems Engineering and Catalysis to Accelerate the Improvements of Chemical Processes

A considerable number of chemical processes nowadays involve some sort of catalytic steps, where catalysts are utilized to convert reactants into intermediates or final products. In some cases, catalysts play a pivotal role in the selection and design of a chemical process, which in turn could lead to significant changes in the economic performance of that process. Therefore, catalysts research and development is of a primary concern to the process systems engineering community, however, there is no systematic approach to identify areas at which catalysts can significantly improve the outcomes of a chemical process. This work is aimed at proposing a systematic approach where catalysis and process systems engineering scientific communities can join efforts in order to accelerate the development of chemical processes. To illustrate an application of the proposed approach, a methanol from shale gas process is taken as a case study. While this process is well-established, the conventional catalyst ($Cu/ZnO/Al_2O_3$) used for the conversion of syngas to methanol takes place under high operating pressure and have low single pass conversion, leading to higher costs associated with the process. A techno-economic assessment is conducted to identify the most influential characteristics of a future methanol synthesis catalyst that could significantly enhance the process's economic performance.

571566 Novel LNG Boil-Off Gas Re-Liquefaction System Design Using Dual-Mixed-Refrigerant Cycle and an Ejector

During transportation of LNG using ships, the fuel is stored cryogenically at 111K and ambient pressure. Due to heat transfer from the environment to the tanks a significant amount of LNG (approximately 0.15% per day) evaporates, known as boil off gas (BOG). This may lead safety issues due the increase in tank pressure, or economic losses if the excess vented without being recovered. However, many methods have been considered, such as combustion, using it as fuel or venting to atmosphere but are undesirable either because of concerns over pollution, low efficiency or high operating cost.

Conceptually for the re-liquefaction system of the BOG is attractive, and have been a number of studies, there are two limitations. In most recent proposed re-liquefaction system, non-liquefied gas remains in the stream returned to the tank, with the surplus gas returned to the tank and still has low efficiency (approximately 41.3%). Thus it was proposed to combine two existing technologies, new dual mixed refrigerants cycle based on cascade cycles and a re-liquefaction system using ejector, and develop it by modifying design. In a novel process design that avoids the return of gas to the tank, while reducing the energy consumption of the compressor stage, to improve energy efficiency. A mathematical model is also introduced for more accurately calculating conditions at the ejector-outlet.

The results show, the power consumption of the compressor stage for BOG entering the system is reduced and performance indicators such as energy efficiency and coefficient of performance, related to operating cost, is much improved than earlier proposals. This study would be the basis of the development of the midstream industry in the aspect of energy consumption and efficiency of the LNG Boil-off gas re-liquefaction system.

578530 Crystallization Process Understanding and Optimization Using FBRM and Raman As PAT Tools

The presentation features the application of process analytical technology (PAT) for understanding the crystallization process in process chemistry development and scale-up. PAT tools such as FBRM, PVM and Raman can be applied as efficient process-monitoring tools to provide real time profile and reliable information much faster than off-line testing. Two challenging case studies will be presented as examples; where PAT tools helped us understand crystallization mechanisms, such as seeding point, nucleation, crystal growth and polymorphic form. This understanding enabled us improve filtration time on scale up by controlling the particle size distribution (PSD) and also optimizing solvent selection for correct polymorphic form. Such information is important for designing science-based process control strategies.

576388 Design and Synthesis of RNA Thermoregulators to Optimize Production of Essential Metabolites and Therapeutic Molecules in Bioprocesses

Bioprocesses nowadays are used to manufacture a wide variety of chemical compounds including polymers, biofuels, and therapeutic molecules. In order to increase the yield of these valuable metabolites, strategies like overexpression and deletion of the involved genes have been implemented. Nevertheless, some of these genes have been found to be responsible for bacterial growth and other essential cell functions, therefore the deletion of these genes would result in more desired product yield per cell but a decreased cell density, and in many cases the result will be cell death. Using genetic switches can be an alternative approach to keep the expression of the involved genes controlled. In this way, the production of the interest metabolite or bacterial growth could be regulated by the metabolic flux. One interesting approach that has been explored in this work is the use of RNA thermoregulators, which are translational controllers of gene expression that respond to temperature shifts by changing their conformational structure. At low temperature, they acquire a secondary structure that precludes the binding of ribosomes to the Shine-Dalgarno (SD) sequence; preventing mRNA translation. Moreover, when temperature increases, RNA thermoregulators reconfigure their conformation, allowing the binding of ribosomes to the SD sequence—which in turn initiates mRNA translation.

Although naturally-occurring RNA thermoregulators are associated with diverse biological processes (e.g. bacterial pathogenesis and life cycle of some bacteriophages), many synthetic analogs have been designed and characterized, for optimal control of gene expression in vivo. In this work, we designed and synthesized different families of synthetic RNA thermoregulators that serve as genetic switches to control different pathways, in Escherichia coli, involved in the production of essential metabolites and molecules with therapeutic properties. This work describes a novel platform technology capable of controlling metabolic flux via RNA thermoregulation in vivo for the production of great value metabolites.

563993 Techno-Economic Analysis of Emerging Vaccine Platform and Production Technologies: How Can We Produce Vaccines Faster and Cheaper?

Conventional vaccine production technologies cannot produce vaccines against new diseases at the rates and volumes required to stop the spread of future epidemics. Additionally, infant and child vaccination demands are often unmet in developing countries due to high upfront vaccine manufacturing facility capital investment costs, inflexibility of manufacturing facilities to produce a wide range of products, and due to cold chain logistic complexities.

Emerging vaccine platform technologies together with advanced bioproduction technologies can empower the production of a wide range of vaccine types, rapidly and at large volumes. Therefore, these exciting new technologies can address both rapid outbreak-response and routine mass vaccination challenges.

However, the vaccine production processes using these new and transformative technologies are not yet established. Furthermore, the total costs of vaccines obtained using these new platform technologies and advanced bioproduction modalities are not well understood. Hence, here process simulation results and the techno-economic analyses of the following 4 new promising vaccine platform technologies is presented: (1) RNA vaccines, (2) Generalized Modules for Membrane Antigen (GMMA), (3) ADDomer vaccines, and (4) humanized yeast-produced vaccines. Additionally, the production costs of these vaccine platform technologies both in conventional and advanced bioproduction facilities are evaluated. For the advanced bioproduction scenario, intensified processes are considered, whereby the facility-related capital and operating expenses can be reduced by containing the production processes inside laminar flow isolators.

Our work indicates that emerging vaccine platform technologies coupled with advanced bioproduction facilities can enable: (a) cost-effective, small-scale manufacturing, promoting out-scalable and distributed manufacturing; (b) rapid and on-demand production; (c) thermostable vaccine products, potentially eliminating cold chain requirements; and (d) flexible production of a wide range of vaccine products using the same production line.

579230 Fused Filament Fabrication (FFF) and Functional Materials

Fused filament fabrication (FFF) of polymers is a 3D additive manufacturing technique that places filaments or tracks one on top of each other to build a three dimensional object. The challenge is two-fold; the weld strength is low resulting in poor mechanical properties and speed of manufacture is dictated by melt flow instabilities such as sharkskin. Sharkskin appears at a critical stress level and is the result of stress relief at the exist of a die. We have used processing aids to reduce sharkskin allowing faster manufacture for poly(lactic acid) or PLA. Contemporary aids are used with polyolefins and these were found to be effective with PLA and in the presentation optimization of their utility will be discussed. Low weld strength can be overcome by thermal annealing of a printed part to induce macromolecular diffusion across the interface, however, this can lead to warping and sagging of the part in general. One could print at higher temperatures to promote diffusion, yet, this leads to a poor part either due to polymeric degradation or the features are simply not as sharp due to sagging during printing. To overcome this limitation, we print two different polymers together, by using two hot ends, simultaneously printing a low temperature polymer (LTP) and a high temperature polymer (HTP). The LTP is the majority of the printed part and constitutes the integrity while the HTP is strategically printed to heat up deposited LTP tracks promoting annealing during the print. This process is called in process thermal annealing and dramatically improves part strength. In addition, a functional material of unique properties is manufactured being a composite of two chemically distinct materials.

570948 Inkjet-Based 3D Printing of Functional Materials

Inkjet printing has evolved from a graphic printing technology to become a fabrication technology for creating functional materials. To date, inkjet printing is capable of depositing a wide range of materials to produce flexible electronics, drug tablets, food, metal and ceramic parts, etc. Many of these emerging applications, however, impose extreme demands on the precise deposition of droplets over large areas. Further, the success of inkjet-based 3D printing hinges on understanding the ink fluid properties and the underlying physics of the print process. A new inkjet tool coined "HuskyJet" has recently been custom-built by integrating a linear sled system with three industrial print heads, in-line UV and IR lamps, and drop imaging. Such design improves the drop-on-drop precision and enables the printing of multi-layer passive circuit elements and various 3D structures. Key lessons learnt from constructing and using this tool will be shared in this presentation.

574936 3D Printed Reconfigurable Liquid Crystal Elastomers Via Light-Triggered Dynamic-Covalent Bonds

3D printing (3DP) of liquid crystal elastomers (LCEs) enables programming of complex, fully reversible 3D shape changes. Classical LCEs must be continuously heated to maintain their programmed actuated shape; here, we demonstrate reconfigurable 3DP LCE architectures by incorporating dynamic-covalent bonds into the LCE backbone. These 3DP LCEs can 'lock-in' on demand to maintain their complex programmed actuated state via bond exchange. Specifically, we create a printable ink containing dynamic-covalent bonds by oligomerizing liquid crystal mesogens with an allyl dithiol chain extender. During printing, the ink is crosslinked to form a LCE that undergoes reversible programmed shape-change upon heating above its nematic-to-isotropic transition temperature, T_{NI}. We leverage light-activated bond exchange to ensure the 'trigger' for bond exchange is orthogonal to the temperature change required for actuation. These reconfigurable LCEs enable architectures to be realized that are inaccessible by printing other classes of soft materials. Finally, by integrating these bond-exchangeable LCEs with standard LCEs via multimaterial 3D printing, we demonstrate a path towards hierarchical, reconfigurable, and responsive structures.

576711 Expanding Membrane Functionality through Additive Manufacturing

Expanding Membrane Functionality Through Additive Manufacturing

William A. Phillip, John R. Hoffman, Feng Gao

Department of Chemical & Biomolecular Engineering

University of Notre Dame

Abstract:

As interest in decentralized water treatment technologies grows, the development of nanofiltration (NF) as a complementary process to reverse osmosis has emerged. However, NF membranes that overcome the tradeoff between selectivity and permeability, that expand the range of molecules that can be selectively separated, and that resist the detrimental effects of fouling must be developed in order to realize the potential of NF-based processes. One approach for overcoming the permeability-selectivity trade-off is through the development of membranes with better controlled pore sizes and pore size distributions. In fact, the performance of NF membranes based on self-assembled materials is already pushing the limits of these size-selective separation mechanisms. Therefore, many researchers are considering approaches that move beyond improved control of size-selective separation mechanisms. In this regard, the functionality lining the pore walls of NF membranes offers another means through which to increase selectivity without sacrificing permeability. Specifically, due to the small pore size of NF membranes, electrostatic interactions and van der Waals forces affect solute permeation dramatically. This feature presents several opportunities for the introduction of functionalities that enable molecular separations based on chemical factors. In regards to fouling, a multitude of antifouling chemistries have been identified, yet introducing these functional chemistries without disrupting the nanostructure of the membranes and in a manner that is consistent with state-of-the-art membrane manufacturing processes remains a challenge.

In this talk, we will discuss how inkjet printing and click chemistry can be used to functionalize membrane chemistries in a precise manner and on time scales consistent with roll-to-roll manufacturing. Specifically, the copper-catalyzed azide-alkyne cycloaddition (CuAAC) reaction mechanism is examined. Using an inkjet printer, reactive ink solutions containing an alkyne-terminated molecule are deposited on nanostructured copolymer membranes with azide-functionalized surfaces. Using Fourier Transform Infrared spectroscopy, the progress of the reaction was quantified by monitoring the decreasing intensity of the peak associated with the azide groups. We will discuss how the rate of membrane functionalization could be controlled through the formulation of the reactive ink solutions as well as the inkjet printing process. In the limit of a rapid CuAAC reaction, full functionalization (i.e., azide conversion) could be accomplished in under 60 seconds. Moreover, a model that predicts the percent conversion of the azide moieties as a function of the reactive solution formulation and processing time was developed and validated. These findings offer a controlled functionalization protocol that is compatible with roll-to-roll processing rates. As such, we provide examples of how this protocol 1) enables the chemical-patterning of membranes in order to access emergent transport mechanisms, and 2) generates multifunctional membranes capable of resisting the detrimental effects of fouling while exhibiting highly selective separations. Through these examples, we will demonstrate that membranes that are functionalized post-assembly using additive manufacturing approaches provide a scalable platform that can be tailored to myriad separations for the treatment and conservation of water resources.

577225 Additive Manufacturing of Mechanically Adaptive Polymer / Nanocellulose Composites

This research highlights materials development and additive manufacturing (AM) of mechanically adaptive polymer composites in filament extrusion processes. Here we specifically focus on the impact of processing on mechanical modulus change upon exposure to water/moisture. A mechanically adaptive composite comprising thermoplastic polyurethane with 10 wt% cellulose nanocrystals is analyzed following 1) filament production from a masterbatch and 2) subsequent AM of functional parts. The interplay between thermal history and shear induced particle orientation are investigated to highlight the magnitude of dynamic modulus change upon moisture exposure, impact of geometry on moisture diffusion kinetics, and reversibility of the adaptation process. Further, we investigate the response of functionally graded AM produced parts to understand the ability to adapt geometry based on stress mismatch.

569495 Material Selection and Formulation for Material Extrusion Additive Manufacturing

Additive manufacturing (AM) has drawn interest from fields ranging from aerospace to regenerative medicine to metamaterials. Using AM, specimens with complex internal geometries and structures can be manufactured. Despite the advantages and interest, broader use of AM is limited by poor mechanical properties, lack of reliability, and lack of expertise. We use experimental and computational approaches to better understand the underlying phenomena that control material extrusion (MatEx) AM. A simulation was developed that describes temperature throughout a build. At benchtop (FFF) scales, short times over T_g were reported, indicating limited opportunity for interlayer diffusion. At larger (BAAM) scales, much longer times above T_g were observed, which can lead to continued flow of extruded material and warping of the printed structure. The effects of material and processing parameters were investigated at both scales, and different trends were observed at the small and large scales. These results help explain experimental observations and provide guidance for material development at all scales of MatEx.

569962 Synthesis of Optimal Hybrid Separation Processes

While distillation operations are the primary vehicles for separation within the chemical industry, they incur high utility costs due to their low thermal efficiencies [1]. Over 50% of energy required for distillation is used for purifying the last 5–10% of the distillate product [2]. This results in significant burden on the economics as well as on the environmental footprint of the overall operation when high-purity end products are targeted. Alternative techniques with membrane and adsorption-based separation technologies require the development of highly selective materials amenable to cost-effective scale up to compete with conventional standalone distillation equipment. Hybrid separation systems combining aspects of both technologies can be optimally designed to reduce energy input and costs while retaining advantages of conventional distillation units. Design of such hybrid systems requires process systems engineering approaches to identify the optimal topology of the separation sequence along with the operational conditions where the synergy is maximum. Determining the optimal topology, e.g. feed location for the distillation column, sequence and connectivity of the different operations involved, etc., in a sequential manner with scenario-based approaches may result in either combinatorial explosion or omission of potential favorable alternatives in an effort to restrict to a tractable number of possible scenarios [3]. Recently, the building block superstructure [4] representation has been demonstrated to provide a systematic methodology for process synthesis and intensification [5-6].

In this work, we extend the use of the building block representation approach for the design and synthesis of hybrid distillation-based separation schemes incorporating membrane and adsorption technologies. For a given design problem, the proposed approach starts with identification of the separation tasks that need to be accomplished. These problems can belong to either completely new process designs or part of an existing flowsheet. Identification of the separation tasks also helps into define targets for improvement. One major target for the hybrid scheme is that all the incorporated separation techniques operate at their highest efficiencies and none of them alone can do the same task better than the hybrid one. After the design targets are set, building block-based superstructure is formed for a hybrid scheme involving distillation, membrane, adsorption, etc. and solved as a synthesis-design-intensification problem addressing the identified separation task and design targets with the optimal design and operating conditions. This is facilitated by solving a mixed integer nonlinear programming (MINLP) problem describing the building block superstructure. To demonstrate the applicability of the proposed approach, optimal membrane-based schemes are synthesized for several hybrid separation schemes intended for solvent recovery applications. An adsorption-based hybrid distillation scheme is also designed for a cryogenic air separation to reduce the high energy input required especially for achieving >95% purities for non-hybrid distillation schemes. The resultant optimal solutions featuring hybrid schemes are examples of intensification that create special hybrid modules for use in processing routes where similar separation tasks exist.

References:

[1] Tian, Y., Demirel, S. E., Hasan, M. M. F., Pistikopoulos, E. N. (2018). An Overview of Process Systems Engineering Approaches for Process Intensification: State of the Art. Chemical Engineering and Processing: Process Intensification, 133, 160-210.

[2] Tula, A.K., Befort, B., Garg, N., Camarda, K.V. and Gani, R., (2017). Sustainable process design & analysis of hybrid separations. *Computers & Chemical Engineering*, *105*, pp.96-104.

[3] Marquardt, W., Kossack, S. and Kraemer, K., (2008). A framework for the systematic design of hybrid separation processes. *Chinese Journal of Chemical Engineering*, *16*(3), 333-342.

[4] Demirel, S. E., Li, J., and Hasan, M. M. F., (2017). Systematic Process Intensification using Building Blocks, *Computers and Chemical Engineering*, 105, 2-38.

[5] Li, J., Demirel, S.E. and Hasan, M.M.F., (2018). Process synthesis using block superstructure with automated flowsheet generation and optimization. *AIChE Journal*, 64(8), 3082-3100.

[6] Demirel, S. E., Li, J., and Hasan, M. M. F., (2019). A General Framework for Process Synthesis, Integration, and Intensification. Industrial & Engineering Chemistry Research. DOI: 10.1021/acs.iecr.8b05961.

569934 Process Intensification Framework for Extractive Separation Systems

The separation of azeotropic mixtures to achieve a higher purity beyond the azeotrope point is a frequent but challenging task in chemical processes. Extractive separation, a classic process intensification invention, offers the solution to overcome the physical equilibrium by introducing an extra mass separating agent (i.e., entrainer) [1-3]. Thus, the entrainers play a key role in these processes due to: (i) their solubility properties to enhance the system's mass transfer driving force to drive separation, and (ii) their hazardous properties (e.g., toxicity, flammability) for environmental and safety considerations.

Recent advances in the synthesis of process intensification systems have been leveraging phenomenological representation methods to systematically generate process solutions without pre-postulation of plausible equipment-based flowsheets which may hinder the discovery of novel design configurations [4-6], which also provides the opportunity to investigate the effect of functional materials at phenomenal level from process fundamental perspective [7]. However, the operational performances (e.g., flexibility, safety, controllability) in the resulting intensified designs, which are crucial to practical implementation, are mostly neglected at this synthesis stage [8]. Therefore, a holistic approach for the synthesis of extractive separation systems, with simultaneous consideration of entrainer selection and operability assessment, is still lacking.

To address this challenge, we have recently developed a systematic framework for the synthesis of operable process intensification systems [9], which features: (i) process synthesis, integration, and intensification using phenomena-based Generalized Modular Representation Framework (GMF) [10] to derive optimal design configurations, (ii) integrated steady-state operability and inherent safety considerations [11], and (iii) simultaneous design and control with explicit model-based predictive control strategies following the PAROC (PARametric Optimization and Control) framework [12]. In this work, we apply the proposed framework to synthesize enthanol/water extractive separation processes with guaranteed operability, safety, and controllability performances. Two entrainer candidates are considered and examined with respect to process profitability, safety, and environmental considerations: a conventional entrainer ethylene glycol (EG) and an ionic liquid entrainer 1-ethyl-3-methyl-imidazolium acetate ([EMIM][OAc]). Mass and/or heat integration are also simultaneously explored to reduce energy consumption, resource utilization, and waste production.

Reference

1. Mahdi, T., Ahmad, A., Nasef, M. M., & Ripin, A. (2015). State-of-the-art technologies for separation of azeotropic mixtures. *Separation & Purification Reviews*, *44*(4), 308-330.
2. Tian, Y., Demirel, S. E., Hasan, M. M. F., & Pistikopoulos, E. N (2018). An Overview of Process Systems Engineering Approaches for Process Intensification: State of the Art. *Chemical Engineering and Processing: Process Intensification*, 133, 160-210.
3. Zhou, T., Song, Z., Zhang, X., Gani, R., & Sundmacher, K. (2019). Optimal Solvent Design for Extractive Distillation Processes: A Multi-objective Optimization based Hierarchical Framework. *Industrial & Engineering Chemistry Research*. **DOI:** 10.1021/acs.iecr.8b04245.
4. Demirel, S. E., Li, J., & Hasan, M. F. (2017). Systematic process intensification using building blocks. *Computers & Chemical Engineering, 105*, 2-38.
5. da Cruz, F. E., & Manousiouthakis, V. I. (2016). Process intensification of reactive separator networks through the IDEAS conceptual framework. *Computers & Chemical Engineering, 105*, 39-55.
6. Tula, A. K., Babi, D. K., Bottlaender, J., Eden, M. R., & Gani, R. (2017). A computer-aided software-tool for sustainable process synthesis-intensification. *Computers & Chemical Engineering, 105*, 74-95.
7. Ismail, S. R., Pistikopoulos, E. N., & Papalexandri, K. P. (1999). Modular representation synthesis framework for homogeneous azeotropic separation. *AIChE journal*, *45*(8), 1701-1720.
8. Baldea, M. (2015). From process integration to process intensification. *Computers & Chemical*

Engineering, *81*, 104-114.
9. Tian, Y., Mannan, M. S., & Pistikopoulos, E. N. (2018). Towards a systematic framework for the synthesis of operable process intensification systems. In *Computer Aided Chemical Engineering*(Vol. 44, pp. 2383-2388). Elsevier.
10. Papalexandri, K. P., & Pistikopoulos, E. N. (1996). Generalized modular representation framework for process synthesis. *AIChE Journal*, *42*(4), 1010-1032.
11. Tian, Y., & Pistikopoulos, E. N. (2019). Synthesis of Operable Process Intensification Systems—Steady-State Design with Safety and Operability Considerations. *Industrial & Engineering Chemistry Research*. **DOI:**10.1021/acs.iecr.8b04389.
12. Pistikopoulos, E. N., Diangelakis, N. A., Oberdieck, R., Papathanasiou, M. M., Nascu, I., & Sun, M. (2015). PAROC – An integrated framework and software platform for the optimisation and advanced model-based control of process systems. *Chemical Engineering Science*, *136*, 115-138.

566916 Process Intensification Via Batch-to-Continuous Transition in the Production of Lubricants

Production of large volume specialty chemicals to-date is conducted almost exclusively in large volume batch or semi-batch reactors. Transition to much smaller continuous processing technology offers the prospect of strong improvements over the century-old batch technology in terms of lower unit energy consumption, smaller footprint, less waste generation, increase in process reliability, and more consistent product quality. The present project, a close collaboration between the University of Pittsburgh and Lubrizol Corporation, aims to demonstrate this transition for succinimide dispersant chemistry as a test case to develop a systematic methodology for batch-to-continuous transition via rapid acquisition of reliable reaction kinetics, implementation in accurate reactor models, validation in a continuous tubular reactor, and, ultimately, construction and demonstration at pilot plant scale. In the present contribution, we focus on validation of reaction kinetics derived from batch experiments in a lab-scale continuous reactor, investigation of the major differences between the two processing technologies and highlight some of the key design considerations for the pilot plant continuous reactor.

The production of succinimide dispersants occurs via amination of a poly-isobutylene succinimide anhydride (PIBSA) with a polyamine. The reaction proceeds via a two-step mechanism: An irreversible amination to an amide acid intermediate, followed by a reversible dehydration reaction to produce water and the desired dispersant molecule. A complete family of dispersants is made by varying molecular weight of PIBSA, structure of the polyamines, and stoichiometric ratio of the two reactants. In the batch process, the water that is co-produced with the desired dispersant product is continuously removed from the head space at the relatively high reaction temperatures, rendering the dehydration step effectively irreversible. In contrast, the presence of water in the continuous reactor renders the dehydration step incomplete due to reversibility and identifies water as a key "contaminant" in this reaction system. However, we find that water removal from the product stream in a subsequent evaporation step results in spontaneous decomposition of the amide acid to the desired dispersant product, resulting in quantitative recovery of the desired product which is chemically and physically indistinguishable from the batch product and fulfills commercial specifications. The presentation will present and discuss these findings and their implications for design of an intensified continuous process.

563339 A Modified Mixed-Acid Catalytic System for Beckmann Rearrangement of Cyclohexanone Oxime in a Microreactor

e-Caprolactam (CPL), the precursor of nylon-6, is mainly produced through Beckmann rearrangement of cyclohexanone oxime catalyzed by oleum in industry. To improve the mixing performance and reaction selectivity of this conventional process, a mixed-acid catalytic system which contains trifluoroacetic acid (TFA) and oleum was firstly proposed and modified in a microchemical system. The effects of experimental parameters on the content of octahydrophenazine (OHP), which are SO_3 concentration, temperature, TFA to oleum ratio (T/A), oleum to oxime ratio (A/O), and TFA distribution in two fluids, were investigated in the microreactor. The content of OHP was reduced with the decrease of temperature and the increase of SO_3 concentration. T/A, and A/O. In addition, a proper TFA distribution is also beneficial for the reaction. A very low content (<1.0 ppm) of OHP was reached under the optimized conditions, which is accord with the requirement of superior-quality CPL.

567421 Fast Pyrolysis of Pine Biomass in a Gas Solid Vortex Reactor

The increased global concern on the dependence on fossil resources has pushed a development in the research of alternative feedstocks for the petrochemical industry. Thermochemical conversion of lignocellulosic biomass into bio-fuels and high valuable chemicals is one of the promising routes to decrease the fossil fuel dependence [1,2]. Fast pyrolysis is one of the most promising thermochemical routes that has been considered for bio-oil production on large scale due to the high bio-oil yields (up to 70 wt. %) [3,4], its simplicity and relatively low capital investment [5]. In spite of the advantages, fast pyrolysis also faces several challenges. The lack of a complete understanding of the pyrolysis reaction network and kinetics makes the tracking of each individual specie (from feed to product) difficult. Yet some simplified kinetic models match the experimental results closely [6]. Another challenge is the selection of the best reactor technology to provide accurate temperature control and fast entrainment and quenching of the pyrolysis vapors (of the order of ms. instead of s).

Nowadays, the fluidized bed reactor (FBR) is one of the preferred technologies for fast pyrolysis due to its ease in design and operation. However, FBRs suffer from operational limits, eg. the maximum fluidization gas flow, while mass and heat transfer rates are moderate. These limits are observed to decrease in a fluidized bed operating in a centrifugal field. The Gas-Solid Vortex Reactor (GSVR) sustains a rotating solids bed by continuous tangential injection of fluidizing gas through multiple small injection slots. GSVRs are known to offer advantages over conventional fluidized bed reactors in the gravitational field, eg. the bed is more densely packed and gas-solid slip velocities are significantly higher. The latter results in enhanced heat and mass transfer rates.

At the Laboratory for Chemical Technology (LCT, Ghent University), biomass fast pyrolysis is carried out in a GSVR demonstration unit [7]. The goal is to further optimize the bio-oil yield and to determine the bio-oil molecular composition using advanced analytical techniques (GCxGC). The present work focuses on experimentally exploring the bio-oil yield in the GSVR demonstration unit. Additionally, CFD simulations including reaction kinetics of this experimental GSVR are performed [8]. By comparing numerical and experimental results, the performance of the selected kinetic model will be validated. First the elemental and molecular composition of the feedstock, and the biomass particle aspect ratio are closely examined. Fast pyrolysis experiments with pine are performed to determine the yield of the main pyrolysis fractions (i.e., bio-oil, char and non-condensable gases). GSVR is compared with conventional FBRs in terms of their respective pyrolysis product fraction distribution. Next, after each experiment, the elemental and molecular composition of the bio-oil obtained from the GSVR is compared with that of bio-oil from FBR technology.

References

[1] Mohan, D., et al. (2006). "Pyrolysis of Wood/Biomass for Bio-oil: A Critical Review." Energy & Fuels 20(3): 848-889.

[2] Czernik, S. and A. V. Bridgwater (2004). "Overview of Applications of Biomass Fast Pyrolysis Oil." Energy & Fuels 18(2): 590-598.

[3] D. Shen, W. Jin, J. Hu, R. Xiao, K. Luo (2015). An overview on fast pyrolysis of the main constituents in lignocellulosic biomass to valued-added chemicals: Structures, pathways and interactions, Renewable Sustainable Energy Rev., 51 761-774.

[4] M.S. Mettler, D.G. Vlachos, P.J. Dauenhauer (2012). Top ten fundamental challenges of biomass pyrolysis for biofuels, Energy Environ. Sci, 5 7797-7809.

[5] Bridgwater, A. V. (2012). "Review of fast pyrolysis of biomass and product upgrading." Biomass and Bioenergy 38: 68-94.

[6] Ranzi, E., et al. (2008). "Chemical Kinetics of Biomass Pyrolysis." Energy & Fuels 22(6): 4292-4300

[7] González Quiroga, Arturo, Pieter Reyniers, Shekhar Kulkarni, Maria del Mar Torregrosa Galindo, Patrice Perreault, Geraldine Heynderickx, Kevin Van Geem, and Guy Marin. (2017). "Design and Cold Flow Testing of a Gas-Solid Vortex Reactor Demonstration Unit for Biomass Fast Pyrolysis." Chemical Engineering Journal 329: 198–210.

[8] Kulkarni, S., Vandewalle, L., González Quiroga, A., Perreault, P., Heynderickx, G., Van Geem, K., & Marin, G. (2018). Computational fluid dynamics-assisted process intensification study for biomass fast pyrolysis in a gas–solid vortex reactor. ENERGY & FUELS, 32(10), 10169–10183.

Cyber-security with Dynamic Watermarking for Process Control Systems

Jaewon Kim, Woo-Hyun Ko, and P. R. Kumar
Electrical & Computer Engineering, Texas A&M University, College Station, TX, U.S.A

Acknowledgment

This material is based upon work partially supported by
NSF Science & Technology Center Grant CCF-0939370,
The U.S. Army Research Office under Contract No. W911NF-18-10331,
The U.S. Army Research Laboratory under Contract No. W911NF-19-2-0033,
and U.S. ONR under Contract No. N00014-18-1-2048.

Abstract

While advances in information and communication technologies have resulted in greater flexibility, scalability, interoperability, and easier administration of process control systems, they have also enabled adversaries to exploit network or software vulnerabilities to subvert sensors or actuators in process control. Adversarial intruders can hack into the system through open networks and conduct malicious attacks to cause critical damage to the entire system. In this paper, we address the problem of cyber-security for process control systems. We specifically address the problem of attacks on sensor measurements. When incorrect sensor measurements are used in feedback control loops, it can lead to incorrect actuation that can cause instabilities or degrade performance.

We investigate a general-purpose solution called "Dynamic Watermarking (DW)" to detect tampering with sensor measurements. It is shown in **[9]** that this technique can be used to secure linear stochastic systems in the presence of malicious sensor nodes or compromised sensor measurements. In **[10]**, it has been shown that DW can detect attacks on autonomous vehicles.

The fundamental idea of the DW mechanism is that actuator nodes superimpose a small private excitation signal on the actuation commands, as shown in Fig. 3. This additive private excitation signal to the "Plant" causes a change in the measurements made Sensor Nodes, in comparison to what they would otherwise be. To detect if the sensor measurements reported are genuine, we design an "Attack Detector", as shown in Fig. 3, that examines if the reported sensor measurements are in conformity with the secret excitation signal. Specifically, it conducts two statistical watermark tests on the reported sensor measurements. It is proved in **[9]** that the Attack Detector can thereby detect malicious activity.

In this paper, we experimentally demonstrate that DW mechanism can be used to detect attacks on sensors in process control systems. We use a representative process control system, a system of two coupled water tanks, as shown in Fig. 1. It has two tanks and a basin. The tanks and the basin are cascaded vertically, so that the upper tank feeds the lower tank whose water flows down into the basin. The control objective is to maintain the water level of both the upper and lower tanks at specified set-points by controlling the voltage of the pump which drives the water from the basin up to the upper tank. The pressure sensors to the controller provide a the level of water in each tank as a feedback at the bottom of each tank.

We experimentally demonstrate the ability of DW by considering two different sensor attacks. The first is a "stealthy" attack. The attacker uses a model of the system dynamics to generate simulated measurements, and reports the simulated measurements rather than the real measurements. The second one is a "replay" attack, where an attacker stores a series of past measurements and replays them later. It is this attack that was reportedly used in the Stuxnet attack **[10]**.

First, we implement each attack and show that each leads to the failure of the water tank control system, causing overflows in the water tanks. Next, we employ DW throughput to secure the coupled water tanks system. It examines the veracity of the reported measurements by checking if they are appropriately correlated with the injected watermark signals. If the statistics of the measurements fail the tests of DW by exceeding predefined thresholds, then the attack detector concludes that the watermarks were distorted or removed, and raises an alarm. What recourse to take once an alarm has been raised generally depends on the particular system at hand. In this case, it stops the entire process for system safety, preventing system failure. We present experimental results showing that the DW mechanism indeed detects the malicious sensor attacks, and thus secures the process control system.

We thoroughly investigate the performance of DW, such as the detection time versus magnitude (i.e., the variance) of the watermark signal, and the parameters of the DW mechanism.

1 Introduction

Figure 1: (a) The Coupled Water-tanks System (b) A Schematic Representation

The particular control system for the two tanks with two sensors and one manipulated variable is shown in Fig. 2a. The general representation of a linear stochastic control system shown in Fig 2b is convenient for describing the dynamic watermarking method. The **Nominal situation**, shown in Fig. 2b is that the honest sensor reports the real measurements $y(k+1)$ to the controller. We model the system as having two types of noises, the process noise $w(k)$, and actuator noise $w_u(k)$.

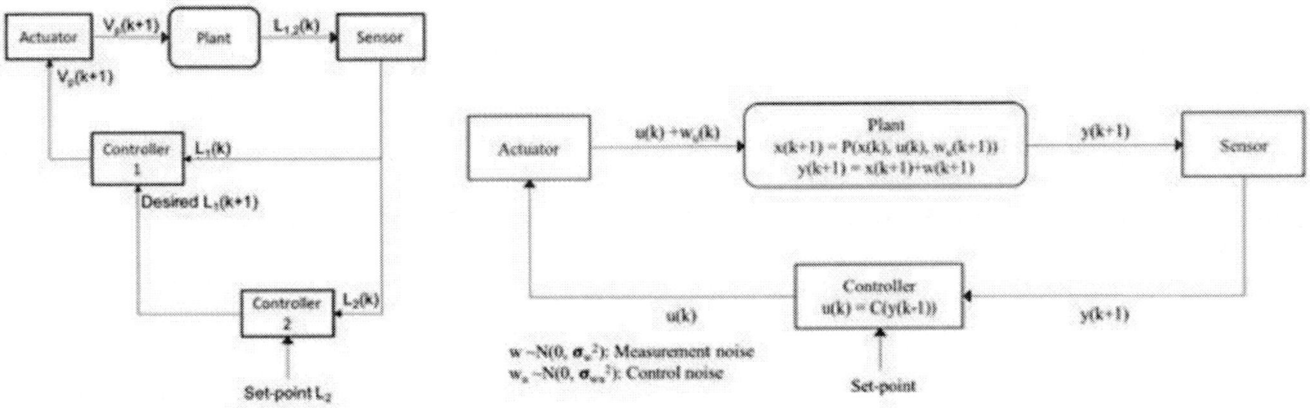

Figure 2: (a) Control System for Coupled Water-tanks (b) A General Linear Stochastic System Representation

2 Dynamic Watermarking (DW)

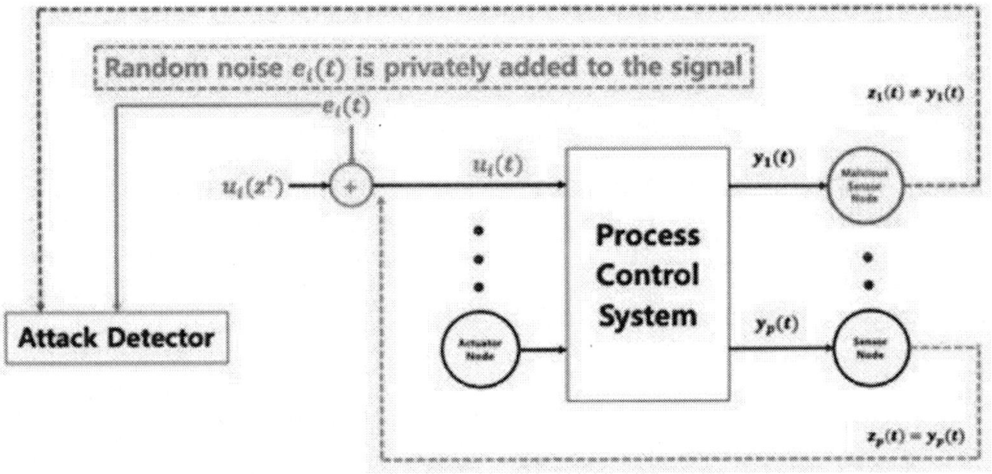

Figure 3: Dynamic Watermarking method for securing cyberphysical systems

The fundamental idea of DW is to have each actuator i add a random signal $e_i[t]$, known as the watermark, on the control policy-specified input $u_i(z^t)$. While the statistics of the watermark e_i are made known to every node in the system, its actual realization is not revealed to any other node $j \neq i$ in the system. Let independent and identically distributed $e[t] \sim \mathcal{N}(0, \sigma_e^2 I)$ denote the vector of watermarks added by all the actuator nodes. The system with output vector $y[t]$, manipulated variable vector u_t, and white gaussian noise $w[t]$ with zero mean and covariance matrix $\sigma_w^2 I$, is

$$y[t+1] = Ay[t] + Bu_t(z^t) + Be[t] + w[t+1]$$

Therefore, the sensor measurements y should satisfy the following equations if the sensor measurements are genuine:

$$(y[t+1] - Ay[t] - Bu_t(z^t)) \sim \mathcal{N}(0,\, BB^T\sigma_e^2 + \sigma_w^2 I)$$

$$E[e_i[t](y[t+1] - Ay[t] - Bu_t(z^t))] = B_i\sigma_e^2,$$

,where B_i = i-th column of B.

Based on this, each honest actuator $i \in 1, 2, ..., m$ subjects the reported sequence of sensor measurements $z(t)$ to the following two T:

TEST 1: The i-th node checks if the reported sequence of measurements satisfy

$$\lim_{T\to\infty} \frac{1}{T} \sum_{k=0}^{T-1} (z[k+1] - Az[k] - Bg_k(z^k))$$

$$(z[k+1] - Az[k] - Bg_k(z^k))^T = \sigma_e^2 BB^T + \sigma_w^2 I_n$$

TEST 2: The i-th node also checks if the reported sequence of measurements satisfy

$$\lim_{T\to\infty} \frac{1}{T} \sum_{k=0}^{T-1} e_i[k](z[k+1] - Az[k] - Bg_k(z^k)) = B_i\sigma_e^2.$$

In **[9]**, it is established that if the reported sequence of measurements passes both Tests, then any malicious sensor present could not have distorted the actual measurement value beyond adding a zero power signal to it.

Experimental Results for Several Types of Attacks

1. Stealth Attack: In this attack, beginning at time 70, the malicious sensor simulates the coupled water-tanks system with randomly generated actuator and measurement noises, and reports the resulting fictitious measurements $z(k+1)$ to the controller.

when watermark signal: mean(μ_e) = 0, var(σ_e^2) = 0.01

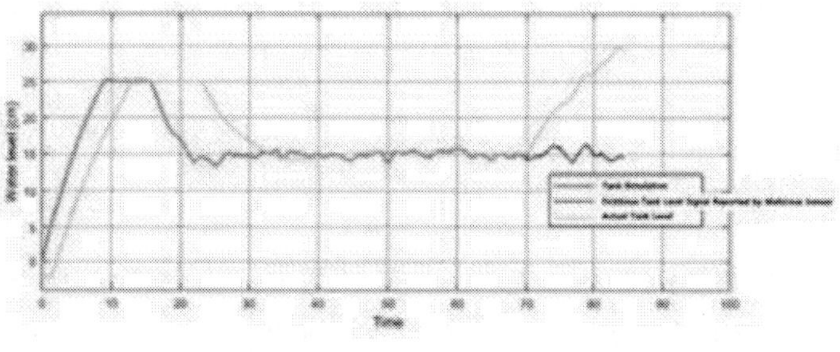

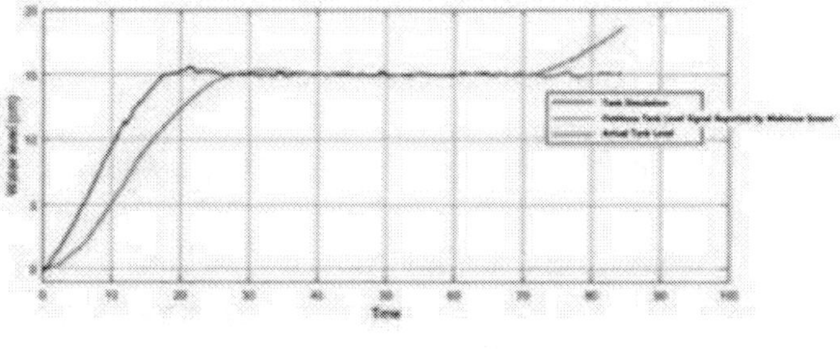

Figure 4: (a) Upper Tank (Tank 1)'s behavior (b) Bottom Tank (Tank 2)'s behavior

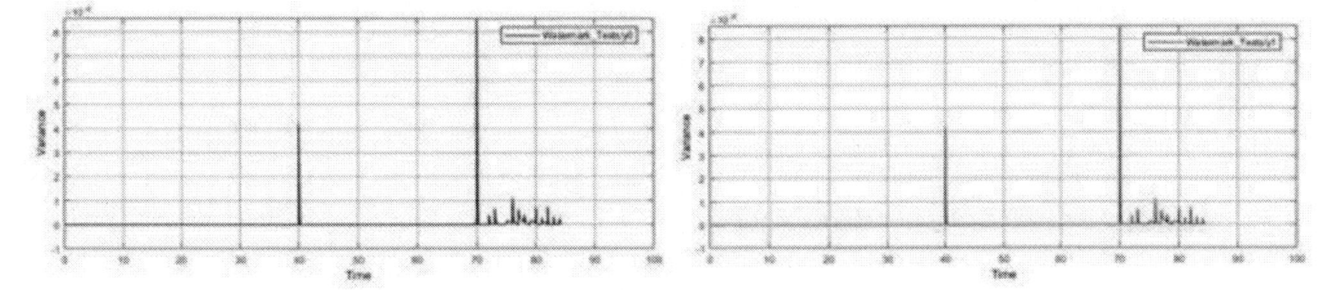

Figure 5: (a) Variance test 1 for Tank 2 (b) Variance test 2 for Tank 2

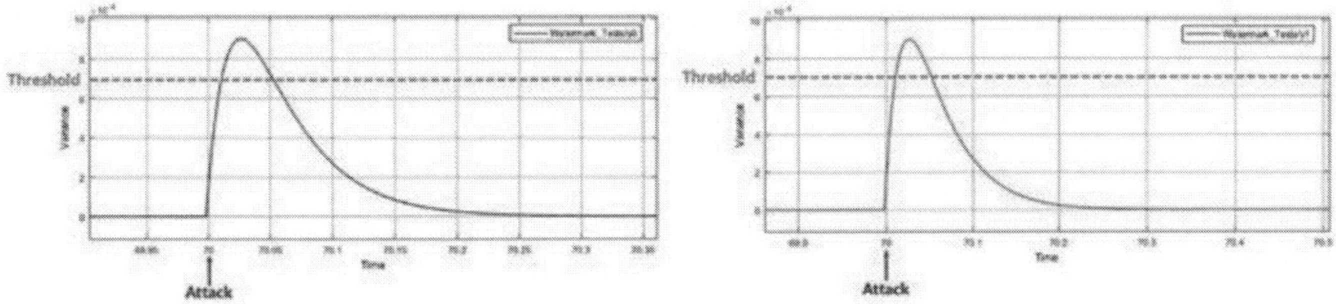

Figure 6: (a) Detection delay of Test 1 for Tank 2 (b) Detection delay of Test 2 for Tank 2

The two dynamic watermarking tests are started from 40 sec. As can be seen in the graph, the attacks result in the tanks are overflowing. However, as can be seen from the Figs. 6&7, the DW tests can detect less than 0.01 seconds after the attack begins. One may note that detection delay is very small even though the watermark signal is of very small variance, just 0.01, so that it does not affect the normal operation of the tanks.

2. Toggling Attack: In this attack, beginning at time 70, the malicious sensor flips the sign of actual tanks measurement values, and reports the resulting fictitious measurements $z(k+1)$ to the controller.

when watermark signal: mean(μ_e) = 0, var(σ_e^2) = 0.01

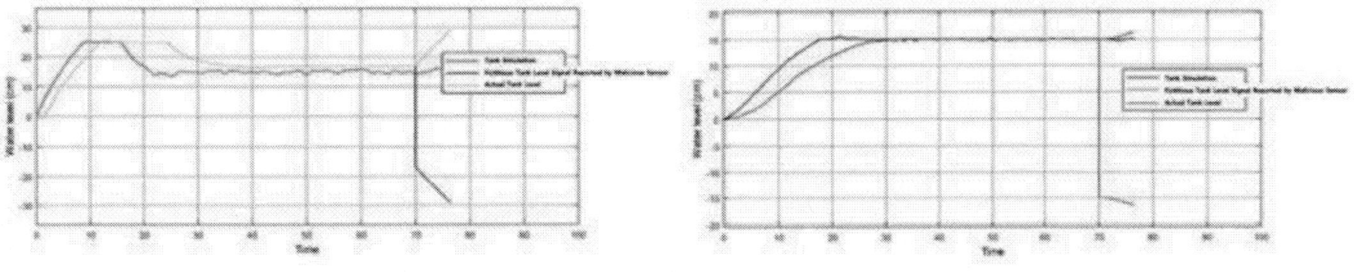

Figure 7: (a) Upper Tank (Tank 1)'s behavior (b) Bottom Tank (Tank 2)'s behavior

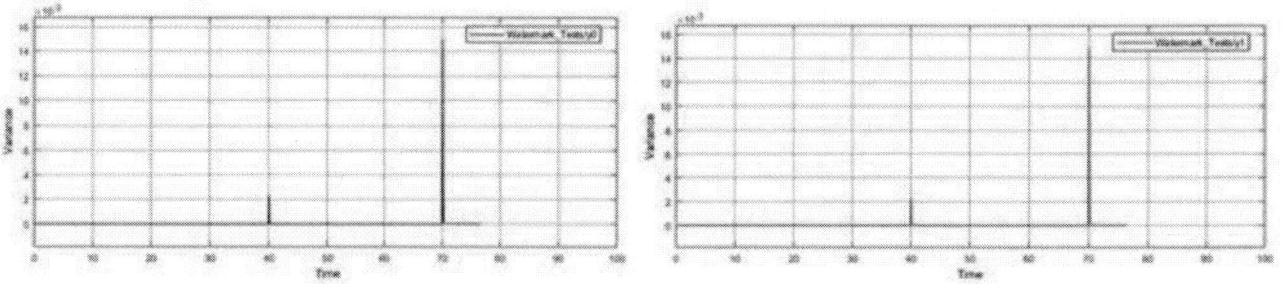

Figure 8: (a) Variance test 1 for Tank 2 (b) Variance test 2 for Tank 2

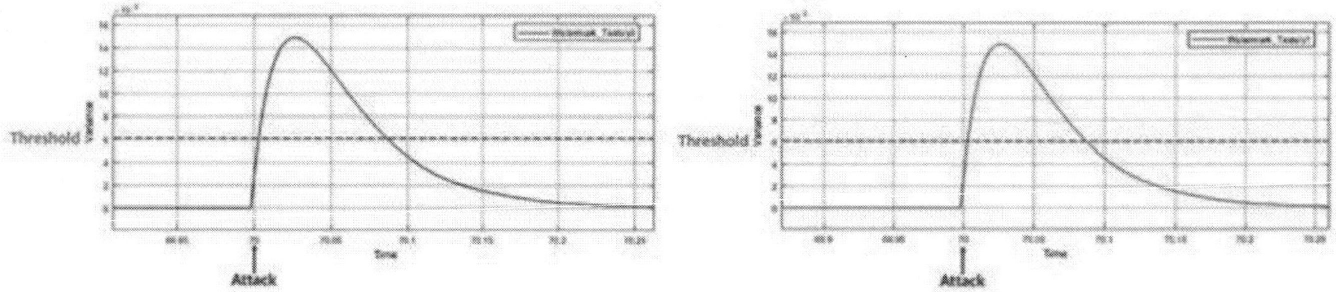

Figure 9: (a) Detection delay of Test 1 for Tank 2 (b) Detection delay of Test 2 for Tank 2

The two dynamic watermarking tests are started from 40 sec. As can be seen in the graph, the attacks result in the tanks are overflowing. However, as can be seen from the Figs. 8&9, the DW tests can detect less than 0.01 seconds after the attack begins. One may note that detection delay is very small even though the watermark signal is of very small variance, just 0.01, so that it does not affect the normal operation of the tanks.

3. Replay Attack: In this attack, from the beginning at time 40, the malicious sensor stores a series of actual measurements and replays them beginning at time 100, and reports the resulting fictitious measurements $z(k+1)$ to the controller.

when watermark signal: mean(μ_e) = 0, var(σ_e^2) = 0.01

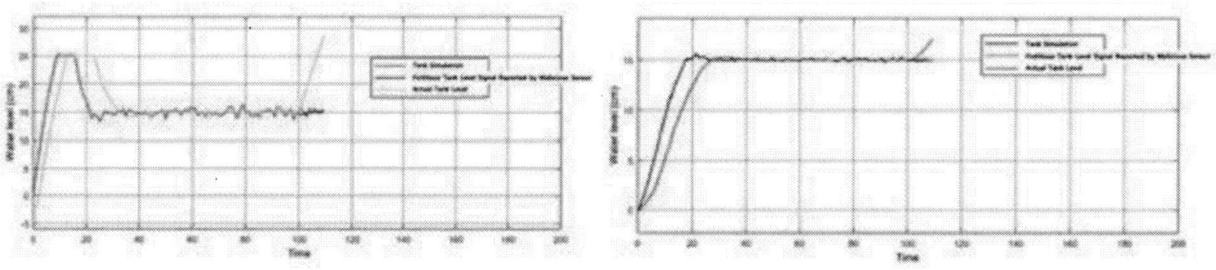

Figure 10: (a) Upper Tank (Tank 1)'s behavior (b) Bottom Tank (Tank 2)'s behavior

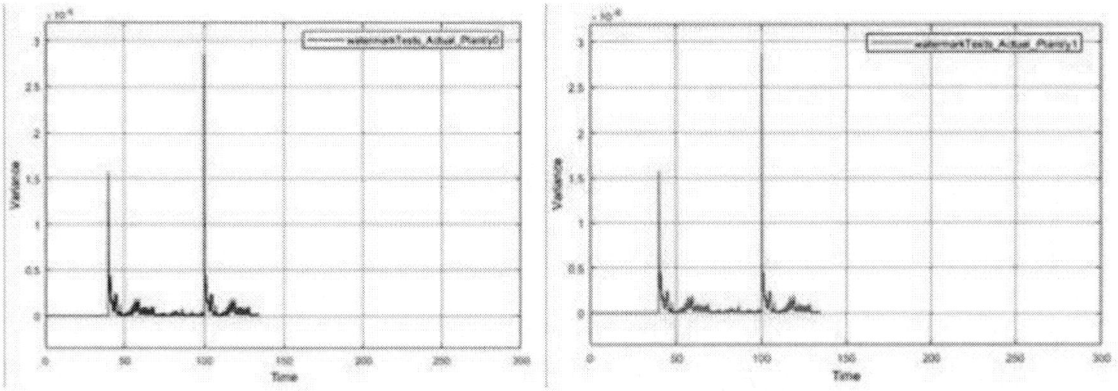

Figure 11: (a) Variance test 1 for Tank 2 (b) Variance test 2 for Tank 2

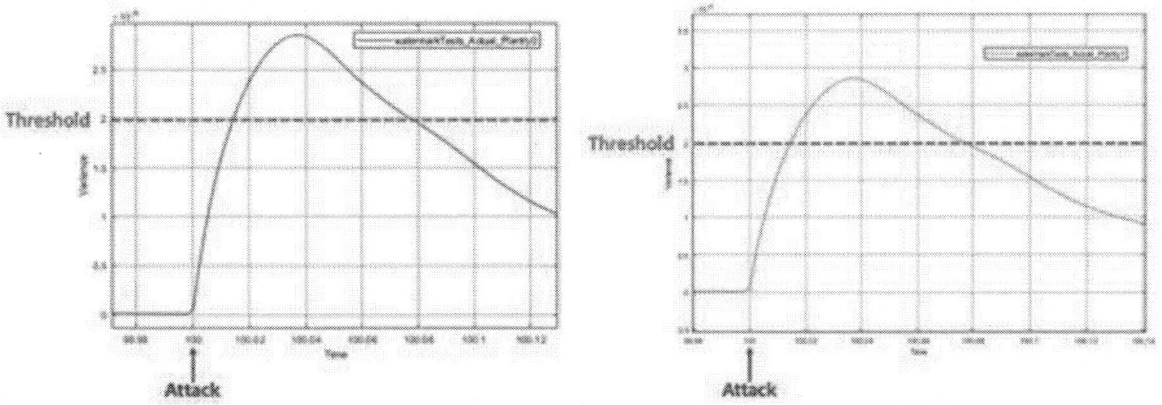

Figure 12: (a) Detection delay of Test 1 for Tank 2 (b) Detection delay of Test 2 for Tank 2

The two dynamic watermarking tests are started from 40 sec. As can be seen in the graph, the attacks result in the tanks are overflowing. However, as can be seen from the Figs. 11&12, the DW tests can detect less than 0.015 seconds after the attack begins. One may note that detection delay is very small even though the watermark signal is of very small variance, just 0.01, so that it does not affect the normal operation of the tanks.

Conclusion

This paper addresses the problem of cyber-security for process control systems. To illustrate the approach and results, we have presented experimental results on the Coupled Water-tanks System. The Dynamic Watermarking method is however a fairly general purpose method and can be used on other systems too. The general idea of dynamic watermarking is to superimpose private excitation signal on the actuation signals. Then by statistically checking whether the reported sensor measurement readings are compatible with the injected random watermark, any malicious tampering with the true sensor measurements can be detected. The advantage of the Dynamic Watermarking approach is that it secures the control system from arbitrary attack strategies. The Dynamic Watermarking method is backward compatible in that it can be implemented on legacy control systems by minimal modifications to the control logic. The effectiveness of Dynamic Watermarking for securing the coupled water-tanks system is indicative of its potential for securing more general process control systems.

References

[1] Ralph Langer. To kill a centrifuge. *The Langer Group, November*, 2013.

[2] Joseph Sangil Kwon, Bharadwaj Satchidanandan, Woo Hyun Ko, Jaewon Kim, Abhinav Narasingam, and P. R. Kumar. Securing process control systems using dynamic watermarking. In *Proceedings of 2018 AIChE (American Institute of Chemical Engineers) Annual Meeting (Computing and Systems Technology Division)*.

[3] Bharadwaj Satchidanandan and PR Kumar. Defending cyber-physical systems from sensor attacks. In *International Conference on Communication Systems and Networks*, pages 150–176. Springer, 2017.

[4] Woo Hyun Ko, Bharadwaj Satchidanandan, and P. R. Kumar. Dynamic watermarking-based defense of transportation cyber-physical systems. *ACM Transactions on Cyber-Physical Systems (TCPS): Special Issue on Transportation Cyber-Physical Systems.*

[5] Bharadwaj Satchidanandan and Panganamala Ramana Kumar. Secure control of networked cyber-physical systems. In *2016 IEEE 55th Conference on Decision and Control (CDC)*, pages 283–289. IEEE, 2016.

[6] Bharadwaj Satchidanandan and Panganamala Ramana Kumar. On minimal tests of sensor veracity for dynamic watermarking-based defense of cyber-physical systems. In *2017 9th International Conference on Communication Systems and Networks (COMSNETS)*, pages 23–30. IEEE, 2017.

[7] Bharadwaj Satchidanandan and PR Kumar. On the design of security-guaranteeing dynamic watermarks. *IEEE Control Systems Letters*, 4(2):307–312, 2019.

[8] Bharadwaj Satchidanandan and PR Kumar. Towards characterizing the watermark-securable subspace of a linear stochastic system. In *Proceedings of 2020 12th International Conference on Communication Systems and Networks (COMSNETS)*. IEEE, 2020.

[9] Bharadwaj Satchidanandan and Panganamala R Kumar. Dynamic watermarking: Active defense of networked cyber–physical systems. *Proceedings of the IEEE*, 105(2):219–240, 2016.

[10] Woo-Hyun Ko, Bharadwaj Satchidanandan, and PR Kumar. Theory and implementation of dynamic watermarking for cybersecurity of advanced transportation systems. In *2016 IEEE Conference on Communications and Network Security (CNS)*, pages 416–420. IEEE, 2016.

575039 Simultaneous Analysis of Control and Cyberattack Detection Algorithms for Enhanced Operational Safety

A critical component of the literature on cybersecurity of industrial control systems has been attack detection techniques. For state measurement falsification attacks, detection methods attempt to identify that an attack is occurring by noting that the measurement data does not match expected trends [1,2,3]. Because sensor measurement falsification cyberattacks make process inputs independent of the actual value of the process state [4], feedback controllers may not be able to prevent cyberattacks from causing harm unless they are notified, via a detection algorithm, that an attack has occurred and that a backup policy must be put in place. One of the drawbacks of detection strategies, however, is that they assume that an attack will take a form that is identifiable as an anomaly by the detection method.

Motivated by the above considerations, we explore the relationship between detection methods and optimization algorithms for chemical processes under model predictive control designs in the presence of state measurement cyberattacks. Specifically, we use a continuous stirred tank reactor example to demonstrate that when the state measurements remain close to the actual values (meaning that the attacks are less "anomalous"), the solutions of the optimization algorithms of the controllers may not differ significantly from the solutions when the state measurement is correct. However, as the state measurements differ more significantly (i.e., a potentially more detectable attack is pursued), the state trajectories can also differ more from the predicted trajectories within the controller. We therefore explore the intersection between the detection thresholds and the optimization problem solution's sensitivity to changes in the state measurement (i.e., initial condition of the process model) to better understand the conditions required for design of cyberattack-resilient detection-control/optimization algorithm combinations.

[1] Z. Wu, F. Albalawi, J. Zhang, Z. Zhang, H. Durand and P. D. Christofides, "Detecting and handling cyber-attacks in model predictive control of chemical processes," *Mathematics*, 6, 173, 22 pages, 2018.

[2] A. A. Cárdenas, S. Amin, Z.-S. Lin, Y.-L. Huang, C.-Y. Huang and S. Sastry, "Attacks against process control systems: Risk assessment, detection, and response," In *Proceedings of the ACM Asia Conference on Computer & Communications Security*, Hong Kong, China, 2011.

[3] I. Kiss, B. Genge, and P. Haller. "A clustering-based approach to detect cyber attacks in process control systems," In *Proceedings of the IEEE International Conference on Industrial Informatics*, Cambridge, UK, pp. 142-148, 2015.

[4] H. Durand, "A nonlinear systems framework for cyberattack prevention for chemical process control systems," *Mathematics*, 6, 169, 44 pages, 2018.

559885 Cybersecurity of Process Control Systems Via a Two-Tier Control Architecture

Since industrial control systems are usually integrated with numerous physical devices, the cybersecurity of control systems plays an important role in safe operation of industrial chemical processes. However, due to the use of a large number of control actuators and measurement sensors and the increasing use of wireless communication, control systems are becoming increasingly vulnerable to cyber-attacks, which is a series of computer actions aiming to compromise the security of control systems (e.g., integrity, stability and safety) and may cause severe industrial incidents [1]. Therefore, the design of advanced detection systems [2] and process control system for nonlinear processes in the presence of cyber-attacks remains an important issue.

In the context of improving process control using a large set of networked sensors and control actuators, a two-tier control architecture was developed for process control systems in [3] to improve plant performance accounting for high-frequency and low-frequency state measurements to handle the increasing number of sensors and actuators. Specifically, a lower-tier control system was developed using PI control based on wired, point-to-point communication with continuous measurements to ensure closed-loop stability, while an upper-tier control system was developed using model predictive control (MPC) based on networked (potentially wireless) communication with asynchronous measurements to improve closed-loop performance. However, in such a two-tier control architecture, the upper-tier control system is more vulnerable to cyber-attacks due to networked measurements. Motivated by this, a secure two-tier control architecture is developed to mitigate the impact of cyber-attacks targeting the upper-tier control system in chemical processes by integrating a machine learning-based detection method [4] with a two-tier control architecture. Specifically, a class of stealthy cyber-attacks that aim to compromise close-loop stability is first designed and applied to the upper-tier control system. Subsequently, a machine learning-based detection method is developed based on continuous state measurements from the lower-tier control system. After the cyber-attack is detected, the upper-tier control system is updated to mitigate the impact of cyber-attacks and re-optimize the closed-loop system. Finally, the effectiveness of the proposed secure two-tier control architecture is demonstrated through a chemical process network example.

[1] Ye N, Zhang Y, Borror C M. Robustness of the Markov-chain model for cyber-attack detection. IEEE Transactions on Reliability, 2004, 53: 116-123.

[2] Ozay M, Esnaola I, Vural F T Y, et al. Machine learning methods for attack detection in the smart grid. IEEE Transactions on Neural Networks and Learning Systems, 2016, 27: 1773-1786.

[3] Liu, J., Munoz de la Pena, D., Ohran, B. J., Christofides, P. D., and Davis, J. F. A two-tier control architecture for nonlinear process systems with continuous/asynchronous feedback. International Journal of Control, 2010, 83, 257-272.

[4] Wu, Z., Albalawi, F., Zhang, J., Zhang, Z., Durand, H., and Christofides, P. D. Detecting and Handling Cyber-attacks in Model Predictive Control of Chemical Processes. Mathematics, 2018, 6, 173.

574665 A Model-Based Approach for the Analysis and Mitigation of Cyber-Attacks in Networked Process Control Systems

With the significant growth in communication and networking capabilities in recent years, and the concomitant advances in sensor and actuator manufacturing technologies, process operations have become increasingly reliant on sensor, actuator and control systems that are accessed over real-time shared communication networks (wired and/or wireless) rather than dedicated, point-to-point links. The integration of a communication network in the sensor-controller and controller-actuator links gives rise to a networked control system and can substantially improve the flexibility and fault-tolerance capabilities of an industrial control system, in addition to reducing the installation, reconfiguration and maintenance time and costs. In addition to the economic and operational benefits of networked control architectures, networked control systems are an important enabling technology for realizing the vision of smart plant operations where advanced cyber-infrastructure and communication technologies are used to tightly integrate, in real-time, process control and operations with business decision making and sustainable environmental, health and safety performance [1].

The increased reliance on networked control systems, while beneficial from a plant performance point of view, poses a host of fundamental challenges due to the inherent limitations on the transmission and processing capabilities of the communication medium. Issues such as network resource constraints, real-time scheduling constraints, data losses and communication delays, may degrade the overall closed-loop performance and even lead to instability if not accounted for in the control system design. This realization has motivated significant research work on the analysis and design of networked control systems, and the literature on this topic is quite extensive (e.g., [2]–[4]).

Another key challenge with networked control systems, especially when dedicated local area control networks are integrated with wireless real-time sensor networks, is the cybersecurity challenge owing to the open nature of wireless networks. Recently, the security of control systems has become the focus of increasing attention due to the deployment of large sensor and actuator networks, as well as the increased use of wireless communication. With these advances, control systems are becoming increasingly vulnerable to cyberattacks, which are a series of computer actions that can compromise the stability and safety of control systems. Cyberattacks generally aim to perturb the process inputs and alter them from what they would have otherwise been under normal operation. Some of these attacks take the form of falsifying sensor measurements sent to the feedback controller, providing incorrect signals to the control actuators, or manipulating stored process data [5].

The communication network and cyber-infrastructure security challenge spans the entire global chemicals industry and petroleum production representing nearly $3 trillion in economic impact. In recent times, cyber-attackers have been successful in causing damage to a uranium enrichment plant, causing power outages, and contaminating a wastewater treatment plant (e.g., [6]-[7]). If left unchecked, these cyber-security risks can potentially lead to physical damage, injury, or death and therefore are a critical problem to address.

Typical approaches for handling control system cyber-security risks are based on computer science, information technology, computer hardware, or networking solutions (e.g., [5], [8]-[10]). More recently, efforts within the process control community have been initiated to address this problem from a control point of view, including the development of frameworks for detecting cyber-attacks and preventing their damage within the context of economic model predictive control [11]-[12].

An approach to reduce the risk of cyber-attacks in a networked control system is to minimize the control system's reliance on the network as much as possible. This idea has been pursued in earlier works (e.g., [13]-[14]) in the context of handling resource constraints. Specifically, a model-based networked plant-wide control structure that enforces closed-loop stability with minimal communication was developed. A set of predictive models were embedded within each local control system and, in conjunction with the local state measurements, the local control action was generated at times when communication between the plant subsystems was suspended, and the states of the models were updated when communication was permitted at discrete times. In doing so, a minimum communication rate that guarantees closed-loop stability could be determined.

While this approach is appealing in that it helps reduce the susceptibility of the control system to network-induced cyber-attacks, it does not provide robustness guarantees against random cyber-attacks. An assessment of the robustness of the networked control system to cyber-attacks is important to identify the fundamental limits within which the system can passively tolerate a cyber-attack, and through this assessment one can also identify the key parameters that can be used to actively mitigate the effects of these attacks when they arise.

Motivated by these considerations, the objective of this contribution is twofold. The first is to develop a framework for the design and analysis of model-based networked process control systems that have well-characterized robustness margins in the presence of certain types of cyber-attacks, and the second is to devise active strategies for mitigating the impact of such attacks. To meet these objectives, we initially construct a model-based plant-wide networked control system in which the local control systems communicate over a shared communication medium. The communication strategy aims to reduce network utilization by embedding predictive models within each local controller and updating the model states by exchanging state measurements over the network periodically at a certain minimum rate.

We consider cyber-attacks in the form of falsified sensor measurements and analyze the impact of using falsified measurements to perform the model state updates at the communication times. By modeling the cyber-attacks explicitly in the closed-loop system formulation, the stability of the networked closed-loop system can be assessed, and an explicit characterization of the stability region in terms of the attack-induced measurement errors, the communication rate and the controller design parameters can be obtained. This characterization then yields the feasible operating range for each system parameter within which robust stability is guaranteed under cyber-attacks, and also reveals the key parameters that can be adjusted in order to mitigate such attacks. Active mitigation strategies are then devised on the basis of the stability region characterization. Finally, the analysis and design results are illustrated using a representative chemical process example.

References:

[1] Christofides PD, Davis JF, El-Farra NH, Clark D, Harris KRD, Gipson JN. Smart Plant Operations: Vision, Progress and Challenges. *AIChE Journal*. 2007; 53(11):2734–2741.

[2] Hespanha JP, Naghshtabrizi P, Xu Y. A survey of recent results in networked control systems. *Proceedings of the IEEE*. 2007; 95:138–162.

[3] Christofides PD, Liu J, Munoz de la Pena D. *Networked and Distributed Predictive Control: Methods and Nonlinear Process Network Applications*. London: Springer-Verlag, 2011.

[4] You KY, Xie LH. Survey of recent progress in networked control systems. *Acta Automatica Sinica*. 2013; 39:101–117.

[5] Khorrami F, Krishnamurthy P, Karri R. Cybersecurity for Control Systems: A Process-Aware Perspective. *IEEE Design and Test*. 2016; 33:75–83.

[6] Langner R. Stuxnet: Dissecting a Cyberwarfare Weapon. *IEEE Security and Privacy*. 2011; 9:49–51.

[7] Clark RM, Panguluri S, Nelson TD, Wyman RP. Protecting drinking water utilities from cyberthreats. *Journal of American Water Works Association*. 2017; 109:50–58.

[8] Pang ZH, Liu GP. Design and Implementation of Secure Networked Predictive Control Systems under Deception Attacks. *IEEE Transactions on Control Systems Technology*. 2011; 20:1334–1342.

[9] Ten CW, Liu CC, Manimaran G. Vulnerability Assessment of Cybersecurity for SCADA Systems. *IEEE Transactions on Power Systems*. 2008; 23:1836–1846.

[10] Linda O, Manic M, McQueen M. Improving Control System Cyber-State Awareness using Known Secure Sensor Measurements. In: *Critical Information Infrastructures Security*, edited by Hämmerli BM, Kalstad Svendsen N, Lopez J. Berlin, Heidelberg: Springer Berlin Heidelberg. 2013; pp. 46–58.

[11] Durand H. A Nonlinear Systems Framework for Cyberattack Prevention for Chemical Process Control Systems. *Mathematics*. 2018; 6(9).URL http://www.mdpi.com/2227-7390/6/9/169

[12] Wu Z, Albalawi F, Zhang J, Zhang Z, Durand H, Christofides PD. Detecting and Handling Cyber-Attacks in Model Predictive Control of Chemical Processes. *Mathematics*. 2018; 6(10). URL http://www.mdpi.com/2227-7390/6/10/173

[13] Sun Y, El-Farra NH. Quasi-Decentralized Model-Based Networked Control of Process

Systems. *Computers & Chemical Engineering*. 2008; 32(9):2016–2029.

[14] Sun Y, El-Farra NH. Resource-Aware Quasi-Decentralized Control of Networked Process Systems over Wireless Sensor Networks. *Chemical Engineering Science*. 2012; 69:93-106.

574448 On the Intersection of Process/Equipment Design and Control System Cybersecurity

Cybersecurity of process control systems is an emerging direction of research investigation in the chemical process control literature. Though a number of works have looked at ways of seeking to combat cyberattacks on industrial control systems (e.g., through detection methods and subsequent strategies which modify signals in a control loop to revert to those which are not compromised by an attack [1,2]), an open research topic in process control system cybersecurity is developing a theoretical framework in which they can be understood. Our recent work [3] demonstrated that for processes described by systems of nonlinear ordinary differential equations, the most general definition of a cyberattack is a condition in which any input within the input bounds can be applied to the process, regardless of the process state. Because the input does not necessarily have a relationship to the state under a cyberattack, these attacks may be difficult to handle with control designs without detection mechanisms that do not fail to identify the attacks before they become problematic.

Motivated by the above considerations, we take an alternative perspective to handling cybersecurity through process control, and instead investigate the manner in which process designs contribute to safety concerns under cyberattacks. Specifically, inspired by the fact that safety relief valves [4] are a mechanical means for preventing unexpected incidents by changing the process dynamics, we explore the manner in which process dynamics and input bounds are related to the extent to which an attack can succeed. We also explore the concept that equipment failure, even that of downstream equipment from the equipment which attacked actuators may directly impact, could be attempted to be compromised during a cyberattack. We explore how modeling of equipment behavior under dynamic process conditions (through equations which describe, for example, stress and strain) may aid in better understanding how cyberattacks can create problematic operating conditions.

[1] Z. Wu, F. Albalawi, J. Zhang, Z. Zhang, H. Durand and P. D. Christofides, "Detecting and Handling Cyber-Attacks in Model Predictive Control of Chemical Processes," *Mathematics*, 6, 173, 22 pages, 2018.

[2] A. A. Cárdenas, S. Amin, Z.-S. Lin, Y.-L. Huang, C.-Y. Huang and S. Sastry, "Attacks against process control systems: Risk assessment, detection, and response," In *Proceedings of the ACM Asia Conference on Computer & Communications Security*, Hong Kong, China, 2011.

[3] H. Durand, "A Nonlinear Systems Framework for Cyberattack Prevention for Chemical Process Control Systems," *Mathematics*, 6, 169, 44 pages, 2018.

[4] T. Marlin, "Operability in Process Design: Achieving Safe, Profitable, and Robust Process Operations." McMaster University, Ontario, Canada, 2012.

569200 Cyber-Physical Security of Advanced Automated Continuous Pharmaceutical Manufacturing Pilot-Plant

Currently, Industry 4.0 concepts are being applied to pharma industry to achieve Pharma 4.0 paradigm. It introduces cyber-physical systems that enable the computerization of pharmaceutical manufacturing. Advanced control, automation, advanced analytics, big data, cloud computing, and internet of things (IoT) are some of the terms commonly referred within Pharma 4.0. It has many advantages but also have bigger challenges on the cyber-physical security side. Over the years, there have been many expensive and notorious incidents in other industry that have prompted the necessity of improved cyber security system. Stuxnet, Duqu, Flame and Shamoon were malware specifically designed to target Industrial Control Systems (ICS) and cause physical damage to the equipment, and inhibit processes from executing normally. Today, process manufacturing in general and the pharmaceutical industry in particular are seeing increasing numbers of cyber security threats. To protect themselves, companies have to change the way they operate by addressing cyber security threats and designing appropriate defenses. Pharmaceutical companies have to recognize their high visibility and attractiveness as targets. They must work to understand cyber security concepts, identify weak points and implement appropriate security measures [1]. Therefore, the cyber security of pharmaceutical manufacturing is highly desired not only economical point of view but also for patient safety perspectives.

The continuous pharmaceutical manufacturing (CM) is one of the most suitable platforms to implement Pharma 4.0 concepts. There are several advantages but also different scientific challenges in implementation and protection of CM [2]. One of the current major challenges for the continuous manufacturing is to protect it from any cyber-physical attack since it has more exposure to system and outside word.

In this work, the cyber security aspects of continuous pharmaceutical manufacturing process have been systematically investigated. The CM pilot-plant is controlled via industry standard control platform. A systematic framework including the methods and tools have been developed for proactive identification and mitigation of potential cyber-physical attacks on CM. A Scalence X208 industrial Ethernet switch (Siemens) has been integrated with the plant so that the network traffic data can be collected. The cyber security relevant software tools such as Snap 7, Wireshark, and Tripwire have been integrated with the CM. The Snap 7 is used to monitor the data blocks to and from the PLC. Wireshark is used for packet sniffing purposes, and the Tripwire is used for the file integrity monitoring. The security data collected from these tools together with the process data collected through HMI (human machine interface) have been utilized to show the behaviors of the system both during normal operation as well as during a simulated attack. Additionally, this work was also focused on distinguishing the malicious attempts from common machine and physical failure.

The objective of this presentation is two folds. First to highlight the developed systematic cyber-security framework and second to demonstrate its application for the cyber-physical security of continuous pharmaceutical manufacturing pilot-plant.

References

1. Lee, C. H. (2018). Boosting Cybersecurity in Pharma. Pharma manufacturing magazine. https://www.pharmamanufacturing.com/articles/2018/boosting-cybersecurity-in-pharma/
2. Bhaskar, A., Barros, F. N., Singh, R. (2017). Development and implementation of an advanced model predictive control system into continuous pharmaceutical tablet compaction process. International Journal of Pharmaceutics, 534 (1-2), 159-178. https://doi.org/10.1016/j.ijpharm.2017.10.003.

562508 3D Printing: A New Era of Precise Manufacturing of Individually Developed Pharmaceuticals

The growing demand for customized medicines increases the impact of 3D printing as one of the emerging technologies for precise manufacturing of individually developed dosage forms. In all 3D printing technologies, large number of materials, geometrical and process factors affect the quality of these computationally designed dosage forms and safety of their use. In addition, for printing an identical dosage form, the quality may vary when built using different 3D printers, even when the same model of the printer, printing parameters, process steps, and raw materials are used. Therefore, knowledge of how the variability of each input parameter and processing step affects the final finished product is critical to ensuring its quality. There is no specific guidance concerning the design, manufacturing process and quality testing considerations for 3D printed solid dosage forms. This presentation will then highlight some of specific methodologies and control strategies considerations over the material, geometrical and process parameters for 3D printing of solid dosage forms. A case study of employing 3D pneumatic assisted microextrusion based 3D printing to manufacture modified release tablets and polypills will be presented. considerations for process validation will be also discussed so that quality for all printlets is maintained when built in a single cycle, between cycles, and by different printers. The issue of identifying and linking the critical factors to the critical quality attributes of the printlets will be also highlighted with an emphasis on the regulatory considerations involved.

561294 Smart Design and Virtual Product Analysis for 3D Printing Pharmaceutical Tablets for Patient Centric Product Development

The Additive Manufacturing is an innovative and promising method that is suitable for the fabrication of personalized dosage forms of tablets. The formulation selection and geometry design of the tablets are key elements for product design. Tuning the formulations and geometry and design of tablet would be based on personalized dose and release profile, and tablets can be smartly designed to provide adequate physical strength and desired drug release profile.

In this work, two mathematical models were developed and validated by experimental data:

The first model is a solid mechanics model by Finite Element Method technique. The model studies the various internal geometry and design of the tablet on tablet axial and radial strength. The sensitivity analysis demonstrates how changing the printing parameters and internal structure of the tablet will impact the tablet hardness. Experimental data for tablet axial and radial strength are in confirmation with the model's results.

The second model is a dissolution and drug release profile simulation. Effects of formulation variabilities (soluble to insoluble ingredients ratio) and tablet geometry on drug release profile have been studied. The model is developed based on unsteady diffusion of water to the solid tablet and swelling (moving boundary PDEs) of the tablet, and unsteady diffusion of drug from the tablet to the dissolution medium.

The models demonstrate how the smart design and virtual product analysis for the 3D printed tablets could be used for patient centric product development.

Disclaimer: This scientific publication reflects the views of the authors and should not be construed to represent FDA's views or policies.

573787 Rapid on-Site Manufacturing Approach with 3DP Capsules: Delayed Burst Release for Regional Absorption

A critical gap existing in our first-to-human strategy is the ability to evaluate complex dosage forms under the on-site formulation (OSF) strategy. We propose to use 3D printing (3DP) to fill this gap by enabling OSF with custom dosage forms using this flexible, high quality, highly reproducible, rapid, and material sparing approach. The 3DP platform has the potential to enable project teams to gain a better understanding of the complex in vitro-in vivo relationships across different drug release mechanisms through accelerated pharmacokinetic screening clinical trials during or immediately following Phase I, and potentially diminish a portion of the decade-long timing from molecule patent to commercialization. Specifically, human regional absorption and permeation is difficult to determine in early clinical phases. The models we use today for calculating regional permeation in the GI tract are largely based on datasets collected in the last few decades using one of two approaches: 1) a double lumen or triple lumen catheter system, or 2) either a single or double tube-balloon system, where the tube would be fed through the intestinal tract and the balloons would be inflated to isolate a portion of the intestines during the study. Both methods are invasive and not feasible as a routine study in humans. An oral delivery system has been developed by Medimetrics called IntelliCap, which is capable of investigating regional absorption; however, this technology is proprietary and therefore may not be cost effective to use for routine studies.

A Hyrel 3D System 30M printer along with various modular heads were purchased from Hyrel 3D, whose heads were modified or custom designed to be able to print new polymer filaments that are semi-brittle, soft, or flexible, or a combination thereof, or to be capable of injecting fluids with viscosities between 20-3000 cP for a single-click print-and-fill dosage form. In this presentation, the development of a set of capsule formulations for delayed burst release with controlled timing is shown to facilitate regional absorption studies on new active pharmaceutical ingredients. By manipulating the thickness of the capsule walls, the onset of release can be controlled. Due to the swelling behavior of PVA, it was discovered that the mechanical properties were critical for robust formulations due to delamination failure that would occur if the layer-layer weld during the printing process was not sufficiently strong. It was also found that fillings containing a large fraction of swellable materials, such as water, affected the physical stability of the formulations. As a first step to enable robust formulations, finite element analysis was used to understand the effect of water-containing drug fillings on the dissolvable capsule shell, and the shell geometry was optimized to reduce swelling stresses that may cause delamination in- vitro or vivo. Second, software techniques were developed, called zoning, to print capsules with no functional defects. Additionally, due to the nature of the 3DP process, in vitro analysis revealed a unique anisotropic dissolution behavior. Custom USP2 hanging baskets were designed and manufactured via 3DP for orientation control during dissolution. Lastly, an in vivo (beagle dogs) study was run using two 3DP capsule formulations and an immediate release compacted tablet as a control, and found a good in vitro – in vivo relationship.

557916 Production of Small-Batches of 3D Printed Multi-Compartment Capsular Devices for Oral Drug Delivery Using High-Throughput Robotic Manufacturing Technologies

In this talk, we present an automated production process enabling the manufacturing of multi-compartment capsular devices for oral drug delivery. This process includes *i)* the production of capsule parts and *ii)* filling the capsules with formulations containing active pharmaceutical ingredients. In this respect, our technology allows fast and efficient fabrication of fixed-dose combination products in small, personalized batches.

The capsular devices manufactured with this process have a structure that is different from the traditional capsules that are already on the market – generally made either of gelatin or hydroxypropyl methylcellulose. Our capsular devices have a modular structure that consists of multiple inner compartments, which are independent of one another. Each compartment can be filled with different active ingredients and with a range of doses, as well as formulations, of the same drug.

Moreover, these capsular devices transform the polymeric barriers of traditionally coated drug delivery systems into self-operating shells. Therefore, they represent a step forward in controlled release – separating the container (i.e. the capsule shell) from the content (i.e. the drug formulations contained in each compartment). The release performance of a compartment is determined by the composition and design features (i.e. morphology and thickness) of its walls. By varying the composition and thickness of each compartment, its release profile can be modified without affecting the properties of the other compartments. In other words, these multi-compartment capsules products yield multiple release kinetics in a single unit.

Multi-compartment capsular devices are especially advantageous for therapies involving multiple drugs that are mutually incompatible or interact in the gastrointestinal tract. Using multiple compartments simplifies the dosing schedule and has a positive impact on overall patient compliance. Moreover, these capsular devices enable the independent development of the drug formulation and the capsule shell, reducing time-to-market and development cost for new drug delivery systems.

The talk will present an example of this new class of capsular devices. The system consists of a two-compartment capsule shell composed of three parts: two hollow halves and a middle part, acting both as a joint and a partition. The joint allows the hollow parts to be assembled into a closed capsular device while dividing the internal cavity into two separate compartments. The hollow parts can have different shapes and thicknesses, resulting in compartments with different release properties and internal volume.

As a proof of concept, we will discuss the fabrication of capsule shells composed by 400 μm-thick and 800 μm-thick compartments using hydroxyl propyl cellulose (Klucel™ LF). We selected 3D printing by fused deposition modeling (FDM) as a suitable manufacturing technique for small-scale batches. This required the development of an industrial-grade FDM system able to realize pharmaceutical products. The talk will discuss in detail the development steps required to manufacture both the filament (through hot melt extrusion) and the capsule shells (through FDM 3D printing).

The FDM process requires as input material thermoplastic polymers supplied in the form of filaments – with suitable diameter and mechanical characteristics. We designed custom extrusion and 3D printing systems, and validated them to fulfill strict pharmaceutical quality requirements. In both cases, the study of the process was aimed at the identification of the critical operating parameters that affect the quality attributes of the final product. We assessed the stability of hydroxyl propyl cellulose after hot-processing and 3D printing, ruling out contamination and the formation of any hazardous degradation product. This led to the formation of an internal protocol for evaluation of by-products. These results demonstrate that the filaments and prototype capsule parts meet the target specifications – in terms of composition, absence of microbiological contamination, and absence of degradation by-products.

Once filled and assembled, the two-compartment capsular devices showed the expected two-pulse release behavior, in agreement with the swellable/erodible nature of the polymer. Furthermore, we compared the dimensional and release characteristics of multi-compartment capsules prepared by fused deposition modeling with those of analogous devices obtained by injection molding (IM). The real-time prototyping capabilities of 3D printing can significantly speed up the R&D process for IM. Both represent viable manufacturing techniques. 3D printing enables the rapid and inexpensive realization of small on-demand batches. Then, once the prototype capsules have been tested and refined, IM allows the large-scale production of capsule shells with the desired behavior.

After the capsule shells are produced, a proprietary robotic system fills each compartment with specific drug formulations. This part of the process relies on a set of automated filling modules working simultaneously, in parallel. The flexibility of this configuration is particularly suitable for the efficient production of small batches of fixed-dose combination products. The automated deposition stage allows the rapid and accurate adjustment of the combination and dosage strength of the active ingredients. Cross-contamination is avoided by enclosing each filling station in a dedicated clean room. A robotic arm (designed to safely collaborate with human operators) moves the capsular devices form one filling station to the next. The capsule filling systems were validated with preliminarily filling trials, which used powders with different particle size distribution and bulk density. We considered 25 batches of 50 capsule each for each powder formulation. In all cases, the amount of active ingredient deposited by the robots had an average coefficient of weight variation (i.e. the ratio between the standard deviation and the average value of the weight of the deposited powder) below 3%. These results demonstrate the accuracy and the flexibility of the powder deposition systems.

This robotic capsule filling system enables the realization of batches that are much smaller than the ones typical of traditional large-scale production processes. This level of personalization requires innovative in-process, non-destructive controls in order to guarantee the quality of the products in a cost-effective manner. We developed quality control machines that are able to check the thickness and weight of the capsule parts. Additionally, we use spectroscopic techniques to check the weight uniformity and identity of the pharmaceutical formulations.

566475 Pharmaceutical Application of Inkjet Based 3D Printing Process

Pharmaceutical Application of Inkjet Based 3D Printing Process

Koyel Sen [1], Sameera Sansare [1], Si Wan Li[2,3], Bodhi Chaudhuri [1,2,3], Anson Ma [2,3]

[1]Department of Pharmaceutical Sciences, University of Connecticut [2]Department of Chemical and Biomolecular Engineering, University of Connecticut [3]Institute of Material Sciences, University of Connecticut

Correspondence: koyel.sen@uconn.edu

Background:

There has been a growing interest in using 3D printing (3DP) for drug tablet manufacturing. This is motivated mainly by 3DP's potential in achieving different dosage geometries, attaining complex drug release kinetics, and combining multiple drugs efficiently for personalized medicine applications. The aims of this study are to: (i) experimentally evaluate an inkjet-based 3D printing method for manufacturing of tablets, (ii) understand the underlying physics, and (iii) to compare against tablets produced by the conventional tableting process.

Methods:

In an inkjet-based 3DP process, a liquid binder is printed onto a layer of excipient powder to bind the powder in a pre-determined pattern. Then, a fresh layer of powder is then spread evenly on top, followed by another round of binder printing. The process is repeated until the construction of the 3D object, or drug tablet in this case, is finished. The surface tension, viscosity, and density of the binder solutions were measured to predict their jettability. This was achieved using dimensionless groups that compare the relative importance of inertia, surface forces, and viscous forces. Two different types of powders and three different types of binder solutions were studied. Successfully printed tablets were further characterized for their weight variation, hardness, disintegration, true density, crystallinity, and physical stability.

Results and Conclusions:

Inkjet-based 3DP was performed using the selected pharmaceutical excipients and binder solutions. The final attributes of the printed tablets were clearly linked to the physical properties of the powder and binder solutions. Such physical understanding was used to create a "design space" in terms of material selection for successful 3D printing of drug tablets. This research lays the foundation for improving the precision and accuracy of the 3DP process for personalized medicine and controllable drug release applications, which will in turn improve patient compliance, lower drug price, and fulfill public health mission.

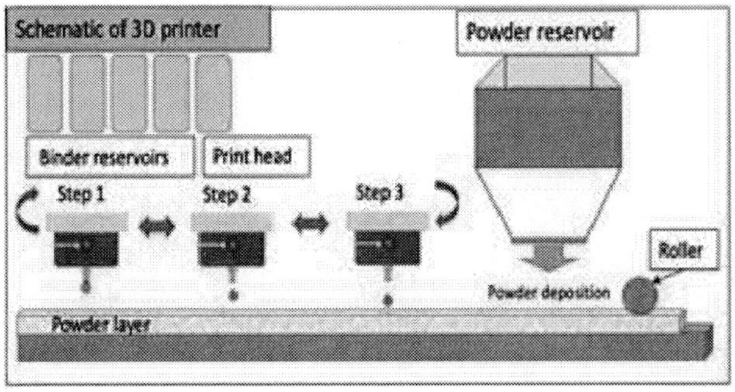

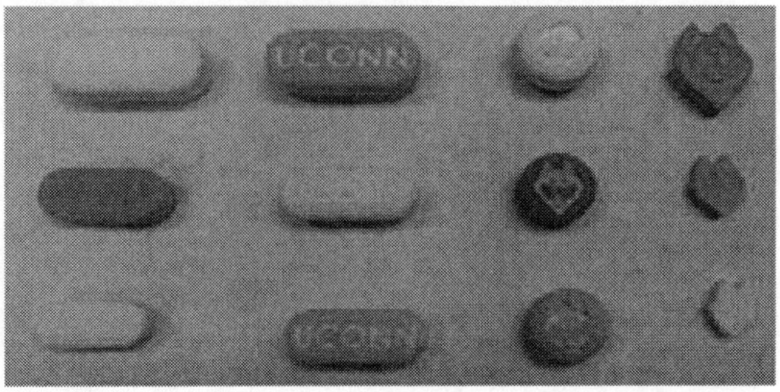

Figure 1- a) Schematic diagram of the inkjet based 3D printer used for the study (left) b) Non-pharmaceutical grade 3D printed tablets using inkjet based printing method

570318 3D Printed Tablets Using Binder Jetting and Pharmaceutical Excipients

Binder jetting is an additive manufacturing (AM) technology used to create three-dimensional model constructs. The materials used in binder jetting are relatively limited generally due to insufficient knowledge surrounding powder technology. The ability to self-formulate powder materials for use in binder jetting machines is not common practice today as not enough rigorous knowledge exists for the requirements of a 3D printable powder.

In this study we characterised several commercial 3D printable powders used in a 3DS Projet460 binder jetting printer, using techniques central to powder flow and powder wetting. The goal was to determine the key powder parameters that are likely to result in a successful binder jetting process, and then to use these parameters to screen standard pharmaceutical excipients for their likelihood to be suitable for 3D printing in a standard commercial printer.

Five different commercial binder jetting powders of various generations from different manufacturers were used as the benchmarks for a binder-jettable powder. Particle size and shape was measured using a Camsizer X2 and the powder flow (bulk and dynamic) was characterised using a Freeman FT4 powder rheometer. All five commercial powders had similar d_{10} and d_{50} values and similar shape distribution profiles. The FT4 showed that all powders fell within a narrow range for the parameters comprising basic flowable energy, specific energy, flow rate index and compressibility.

Drop penetration tests were used to determine the likely construct quality prior to the powder being 3D printed. The dried nuclei from these tests were analysed for spread, depth and hardness, which correlates to the dimensional accuracy of the powder after 3D printing. Drop penetration time seemed to have a wide range of acceptable values, but nuclei construct strength was important. A preliminary framework highlighting the quantitative parameters for binder jetting powders was created.

Finally, to test the framework approach, a simple formulation of 20% VA64 Kollidon and 80% Microcrystalline cellulose was used to successfully 3D print a range of basic tablet shapes in a Projet460 binder jet printer. The use of these commercial powders and characterisation techniques ultimately provide a useful benchmark for determining the suitability of standard pharmaceutical powder APIs and excipients to achieve good quality printed constructs, and sets out the considerations for specially engineered pharmaceutical particles and formulations that would need to be developed in future.

575111 Potential of 3D Printing in Precision Combination Therapy: A Case Study of Cardiovascular Polypills

3D printing is one of the emerging manufacturing technologies that is well suited to produce precision (personalized) medicine. The unique versatility of 3D printing to engineer desired designs with specific geometry allows for achieving targeted release profiles of printed drugs. It has a high potential for individualization of drug therapy combination through the so called "Polypill printlets" by which all the drugs required for the therapy can be printed into a single dosage unit. In this regard, this presentation will highlight a case study of engineering and manufacturing a cardiovascular polypill of Metoprolol (MT) and Atorvastatin (AT) using 3D pneumatic assisted micro-extrusion (PAM) technique. The polypills were designed to provide extended release properties of MT and immediate release properties of AT. The effect of printing, geometrical and formulation parameters on breaking strength and disintegration of the polypills and release properties of the drugs will be discussed. Various designs of the polypills will be demonstrated to modulate the release rate of the drugs from the polypills. The design spaces for formulating these polypills will be discussed in terms of identifying the critical material attributes and key geometrical parameters that achieve independent release properties of both drugs from the polypills. In addition, the regulatory considerations and failure modes of the polypill printlets will be also shared.

DISCLAIMER

This abstract reflects the views of the presenter and should not be construed to represent FDA's views or policies.

574353 Utility Functions for Bayesian Design of Tests for Fault Detection and Isolation

Modern cyber-physical systems (CPS) are often characterized by complexity and corresponding parametric uncertainty. The latter necessitates advanced fault detection and isolation (FDI) algorithms that can handle the instantiation of uncertainty as false alarms or non-detections.[1] The implementation of experimental design for an active FDI can be an important asset when it is feasible. The approaches[2–4] cast active FDI as an exercise of design of experiments that generates tests to maximize the information extracted from available system sensors. However, model-based active FDI approaches are not popular in the literature due to their vulnerability to parametric uncertainty and model error, which result in non-robust designs. Bayesian design of experiments (BDoE) for the design of FDI tests is capable of maximizing test information for the entire model/parameter space and provide a robust solution.[5]

Bayesian approaches are based on the Bayes theorem, where a posterior belief on parameters can be updated using prior information on parameters multiplied with the likelihood of the observed data. In BDoE, an optimal design can be found by maximizing the expected utility based on the observed data. BDoE can utilize a variety of metrics as utility function, such as the Fisher Information Matrix (FIM) or Shannon Entropy of Information.[5] The choice of the utility function impacts the solution robustness and the computational cost of the BDoE. For instance, adopting FIM as a utility function can lead to an alphabetical design criterion such as D- or D_s-optimality. However, these criteria involve sensitivity analysis to deterministically calculate the FIM of the system and consequently are computationally expensive. On the other hand, when Kullback-Leibler divergence from prior to posterior is used to make an inference on parameters, the expected utility calculations involve a nested Monte Carlo estimation which affect both computation time and prediction.[6]

Our discussion will begin with a formalistic comparison of Frequentist (classic) experimental design and BDoE for different objective metrics. We will focus on the impact that different utility functions have on the optimal design and their computation time. To illustrate the importance of BDoE, a variation of the benchmark 3-tank system (Figure 1) is used as a case study, while we will present the deterministic variance of the optimal design for different utility functions for this case study.[7] We will show how different utility functions can be used for inferences on parameters for FDI, depending on the data availability and the associated computational cost.

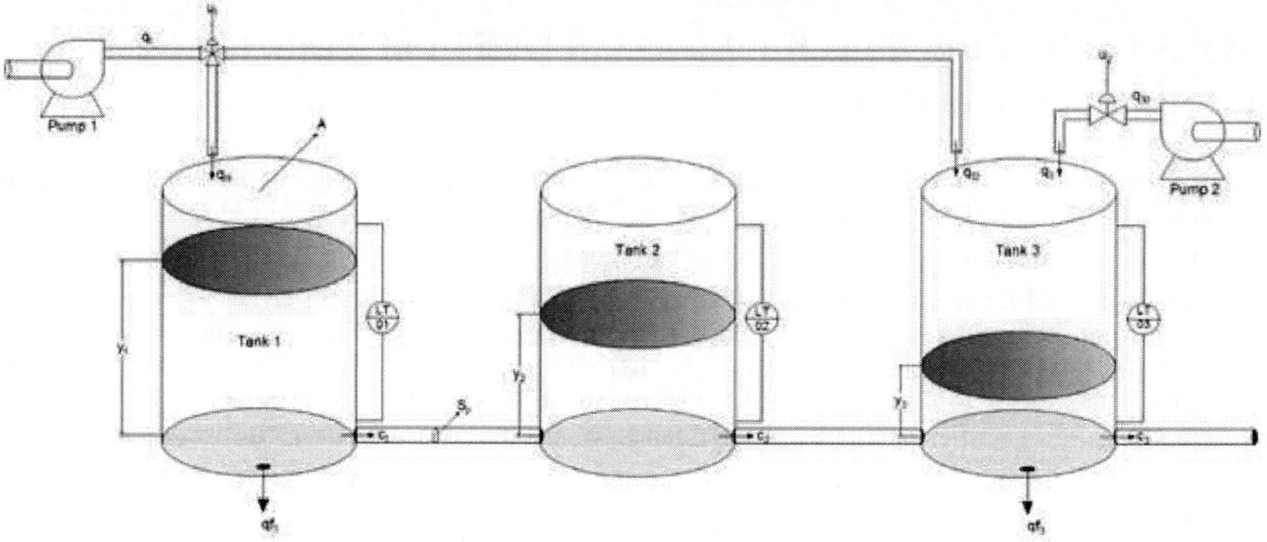

Figure 1: Three Tank System

Acknowledgment

This work was sponsored by the UTC Institute for Advanced Systems Engineering (UTC-IASE) of the University of Connecticut and the United Technologies Corporation. Any opinions expressed herein are those of the authors and do not represent those of the sponsor.

References

1. Venkatasubramanian, V., Rengaswamy, R., Yin, K. & Kavuri, S. N. A review of process fault detection and diagnosis: Part I: Quantitative model-based methods. *Comput. Chem. Eng.* **27**, 293–311 (2003).
2. Palmer, K. A., Hale, W. T., Such, K. D., Shea, B. R. & Bollas, G. M. Optimal design of tests for heat exchanger fouling identification. *Appl. Therm. Eng.* **95**, 382–393 (2016).
3. Palmer, K. A. & Bollas, G. M. Active fault diagnosis for uncertain systems using optimal test designs and detection through classification. *ISA Trans.* (2019). doi:10.1016/j.isatra.2019.02.034
4. Palmer, K. A., Hale, W. T. & Bollas, G. M. Active Fault Identification by Optimization of Test Designs. *IEEE Trans. Control Syst. Technol.* 1–15 (2018). doi:10.1109/TCST.2018.2867996
5. Chaloner, K. & Verdinelli, I. Bayesian experimental design: A review. *Stat. Sci.* 273–304 (1995).
6. Huan, X. & Marzouk, Y. M. Simulation-based optimal Bayesian experimental design for nonlinear systems. *J. Comput. Phys.* **232**, 288–317 (2013).
7. Mesbah, A., Streif, S., Findeisen, R. & Braatz, R. D. *Active fault diagnosis for nonlinear systems with probabilistic uncertainties. IFAC Proceedings Volumes (IFAC-PapersOnline)* **19**, (IFAC, 2014).

575087 Nonlinear Observer Design for Process Monitoring in the Presence of Fast and Slow Modes

Sensors are widely used in chemical processes for safe operation and product quality monitoring purposes. However, in many practical applications, only some of the variables critical for quality control are available for measurement. This is particularly significant in bioreactor systems, where some important species concentrations are either unavailable or only available with significant delay and high cost, due to the lack of appropriate online estimating devices or the high price of sensors. In these circumstances, state observers play an important role in process monitoring.

Processes with multi-time-scale dynamics are common in chemical engineering. In fact, many bioreactor systems involve both fast and slow dynamics owing to the presence of multiple microbial cultures.

These systems could pose challenges to process simulation due to stiffness issues, and are thus computationally costly for optimization. To overcome this, one option is to do model reduction and properly eliminate the fast modes. [1]

It is also a challenging problem to design observers for systems showing spectral gap. On one hand, there may be ill conditioning problems, and the selection and placement of observer poles could have a critical effect. On the other hand, high gain observers may perform well, but could result in severe peaking problems and are likely to fail in the presence of measurement noises. Many researchers have studied observer design problems for linear multi-time-scale systems. In recent work by Yoo at al.[2], the authors decomposed a singular perturbed system into fast and slow subsystems through a series of linear transformations, and placed the observer poles separately for the two subsystems. The overall observer shows less peaking problem and converges faster than the traditional observer in their numerical example. For nonlinear systems, nonlinear observers generally show better stability and performance properties, as the linear observer is designed only for local validity. But in the presence of spectral gap, the design problem is even harder than in the linear case. Kazantzis et al.[3] has proposed an approach to design observers for slow states in singular perturbed systems, by first designing observers for the corresponding reduced model and then applying it back to the detailed model. Inspired by these works, we will consider more general nonlinear multi-time-scale systems and propose two approaches to design nonlinear observers for them.

In the presentation, we will explore two alternative design approaches. The first approach is to introduce a systematic way to properly select the design parameters for the observer PDEs [4] for the detailed model. The first step in this approach will be to design a linear observer for the linearized model and use the matrices of the resulting error dynamics as the design parameters for the nonlinear observer PDE. The second approach is to combine observer design with model reduction using the invariant manifold method [1]. Similarly to the work of Kazantzis et al. [3], a nonlinear observer designed for reduced model will be applied on the detailed model. We will show that this observer derived from the reduced system will also converge for the original detailed model.

The theoretical methods will be applied to two bioreactor case studies. In the first case study, the aerobic digestion system involves three states: the concentrations of biomass and substrate, and the dissolved oxygen tension (DOT). With the measurement of biomass concentration and DOT, we will estimate the substrate concentration. In this system, DOT dynamics is generally much faster than the other two. The second approach would then be based on a reduced model without DOT.

The second case study is for an anaerobic digestion system, with the concentrations of substrate, VFA, acidogens and methanogens to be the four states. By measuring the total biomass concentration and the VFA concentration, the unmeasured three states are to be estimated. Stamatelatou et al.[5] has studied the properties of this system and shown that the substrate concentration is associated with much faster dynamics than the other three. In this case, the reduced model derived by Stamatelatou et al.[5] will be used for the second approach. We will also compare the two methods and evaluate their performance on the case studies.

References

[1] Kazantzis, N., Kravaris, C. and Syrou, L., 2010. A new model reduction method for nonlinear dynamical systems. Nonlinear Dynamics, 59(1-2), p.183.

[2] Yoo, H. and Gajic, Z., 2018. New designs of linear observers and observer-based controllers for singularly perturbed linear systems. IEEE Transactions on Automatic Control, 63(11), pp.3904-3911.

[3] Kazantzis, N., Huynh, N. and Wright, R.A., 2005. Nonlinear observer design for the slow states of a singularly perturbed system. Computers & Chemical Engineering, 29(4), pp.797-806.

[4] Kazantzis, N. and Kravaris, C., 1998. Nonlinear observer design using Lyapunov's auxiliary theorem. Systems & Control Letters, 34(5), pp.241-247.

[5] Stamatelatou, K., Syrou, L., Kravaris, C. and Lyberatos, G., 2009. An invariant manifold approach for CSTR model reduction in the presence of multi-step biochemical reaction schemes. Application to anaerobic digestion. Chemical Engineering Journal, 150(2-3), pp.462-475.

575451 Model-Predictive Safety: Min-Max Optimization to Calculate the Most Aggressive Control Actions and the Worst-Case Uncertainties

In 2016 [1] we introduced a method of model-predictive safety (MPS) system design, and in 2017 and 2018 [2, 3] min-max optimization problems that should be solved offline to implement an MPS system in real time. The solutions of the optimization are (a) the most aggressive control action that minimizes each process-constraint index when uncertain model parameters take their nominal values, and (b) the most aggressive control action that minimizes each process-constraint index when uncertain model parameters take their worst-case values.

An MPS system [1] generates alarm signals that are predictive and systematically account for process nonlinearities and interactions, while typical existing functional safety systems generate reactive, non-interacting alarm signal(s) when a process variable exceeds a threshold. The MPS system design method allows for a systematic utilization of dynamic process models to generate alarm signals (alerts) for the predictive detection and proactive prevention of operation hazards (OHs) in real time. An MPS system uses a process model to predict the process safety status over a moving prediction horizon and to generate alarm signal(s) indicating the presence of a present or future OH with reasonable accuracy; it generates alarm signals that alert the process personnel to imminent and potential future OHs before the actual OHs occur.

In this paper, we propose and implement a particle-swarm optimization (PSO) method to solve the min-max optimization problems described in Ref. [2] for two process examples, a classical chemical reactor with series reactions and a free-radical polymerization reactor. Results from these two examples show that the PSO method reliably finds offline (a) the most aggressive control action that minimizes the infinity norm of each safety-constraint index over a prediction horizon, and (b) the most aggressive control action that minimizes the uncertainty-maximized infinity norm of each safety-constraint index over a prediction horizon. The performances of the MPS systems implemented with different prediction horizons are shown using numerical simulations.

References

[1] Mohseni Ahooyi, T., J.E. Arbogast, W.D. Seider, U.G. Oktem, and M. Soroush, "Model-Predictive Safety System for Proactive Detection of Operation Hazards," AIChE J., 62, 2024-2042 (2016).

[2] Soroush, M., J.E. Arbogast, and W.D. Seider, "Model-Predictive Safety System for Predictive Detection of Operation Hazards: Off-Line Calculation of Most Aggressive Control Actions and Worst-Case Uncertainties," CAST Division 10 Plenary Session at the 2017 AIChE Annual Meeting, Minneapolis, MN (2017)

[3] Soroush, M., A.A. Shamsabadi, W.D. Seider, and J.E. Arbogast, "Implementation of Model-Predictive Safety Systems to Detect Predictively Operation Hazards in Non-Minimum-Phase Processes," 2018 AIChE Annual Meeting, Pittsburgh, PA (2018)

576766 Exploring the Role of System Knowledge in Data Analytics Using an Iiot-Enabled Smart Manufacturing Testbed

The emergence of the industrial Internet of Things (IoT) and ever advancing computing and communication technologies have fueled a new industrial revolution. This revolution is happening worldwide to make current manufacturing systems smarter, safer, and more efficient. Although many general frameworks have been proposed for IoT enabled systems and their potential in industrial applications, there is limited literature on demonstrations or testbeds of such systems. Simulation is a powerful tool but the fidelity of the simulated system is limited by the understanding on the system; that is, the model that describes the system. With industrial IoT still in its infancy, there is not sufficient understanding on the property, capacity and performance of IoT sensors to enable accurate simulation. In addition, there is a lack of systematic study on the characterization and representation of data generated by industrial IoT sensors and data analytics challenges associated with industrial IoT sensor data. These are the major motivations for us to build a smart manufacturing testbed equipped with IoT vibration sensors, and to investigate the capabilities of these IoT vibration sensors for process monitoring.

In this work, we will discuss the details of the testbed setup and the characteristics of the data collected from the IoT vibration sensors. We investigate strategies to address IoT sensors' data challenges and their impacts on data analytics. We show how data visualization and exploration can help identify underlying process patterns and correlations. We demonstrate how data analytics performance and reliability, as well as results interpretability can be significantly enhanced by incorporating system knowledge with data analytics techniques.

577053 The Nature of Fault-Space Classification By Neural Networks

Neural networks have emerged as a widely adopted tool used for process monitoring, fault diagnosis, and control. However, a major drawback is their inability to provide mechanistic explanations and insights about their decisions. In this paper, we report on a systematic approach to understand how a neural network performs classification by characterizing the structure of its feature space. Using carefully chosen model problems, we probe the structure of the feature space under various controlled conditions to elicit key insights. We show, for example, under what conditions adversarial attacks might succeed. We propose techniques to couple a neural network's learning with input modeling to make the system more robust in a classification task.

571285 Design and Optimization of 3D Reactor Technologies for the Production of Light Olefins

Roughly 1.5 x 10^8 t of ethylene and 8 x 10^7 t of propylene are produced annually worldwide. Although some alternative technologies have emerged and/or gained interest over the last decade, the bulk of these molecules are and will continue to be produced through the steam cracking process. [1]

In this process, a hydrocarbon feedstock is heated together with steam in a tubular reactor to temperatures of up to 900 °C. To reach these high temperatures, the reactors are typically suspended in a gas-fired furnace. Due to these elevated temperatures and large scale of the process, it is one of the most energy-intensive processes in chemical industry, accounting for an annual energy consumption of about 3.0 x 10^{18} J, [2] which is the equivalent of approximately 500 million barrels of oil or twice the annual electricity consumption of Germany. Next to this, the process is also responsible for the emission of nearly 200 million tons of CO_2.

As the process has already been established for a long time, a lot of effort has been put into increasing the process efficiency, leading to a high level of complexity and limited room for further process improvements with conventional optimization techniques. In this respect, Djokic et al. [3] propose technological innovations to push beyond contemporary conventional steam crackers. These innovations range from alternative fuels such as oxy-fuel over material improvements such as high-emissivity coatings or advanced high temperature alloys to the enhancement of convective heat transfer from the reactor coil to the process gas by implementing 3D reactor technology.

These 3D reactor technologies not only reduce the fuel consumption and lead to longer coil lifetime by reducing the tube metal temperature (TMT), they also decrease coke formation which increases the process run length. This is mainly achieved by two methods: on one hand, the reactor wall shape can be altered to increase the heat exchanging surface. On the other hand, inserts or wall modifications can be introduced to increase the degree of mixing. This can be achieved at a large scale by radial mixing, i.e. secondary motions displacing hot fluids away from the reactor wall into the colder center of the flow, and on a smaller scale by increasing the turbulence in the near-wall region. This near-wall turbulence increase, will decrease the thermal boundary layer, which is the region in which the largest resistance to heat transfer can be located.

Even though a large number of heat transfer enhancing designs have been proposed and/or implemented in the field of heat exchangers, the number of designs suited for high temperature applications such as steam cracking is rather limited. Several reasons can be mentioned for the limited range of designs; specialized coil materials do not easily allow complex deformations or 3D patterns; turbulators can increase the risk of coke depositions clogging the reactor and the use of these designs inevitably leads to an increase in pressure drop over the reactor, which might negatively affect the yield of the desired light olefins.

Comparisons between different designs in an industrial setting are not widely available. To cope with this, flow, heat transfer and chemical kinetics are now modeled by means of Computational Fluid Dynamics, providing a unique insight into the effect of reactor design on flow behavior and heat transfer. This modelling effort is validated based on the results of an extensive experimental campaign, where different enhancement techniques are investigated with stereo-PIV, for flow field measurements, as well as heat transfer measurements, based on liquid crystal thermography (LCT). These experiments provide detailed local information and hence allow a thorough comparison with the simulated flow field and heat transfer capabilities of these designs.

The modelling methodology is used to investigate the performance of two recently developed reactor technologies namely SFT[4, 5] and SCOPE[6]. These technologies are then compared against three established reactor designs; finned[7] and ribbed[8] coils as well as coils with IHT inserts[9]. Simulation results will focus on pressure drop, tube metal temperatures, product distribution and coking rates, allowing a unique comparison of the most relevant industrial reactor designs.

[1] I. Amghizar, L. A. Vandewalle, K. M. Van Geem, and G. B. Marin, "New Trends in Olefin Production," *Engineering,* vol. 3, pp. 171-178, Apr 2017.

[2] T. Ren, M. K. Patel, and K. Blok, "Steam cracking and methane to olefins: Energy use, CO2 emissions and production costs," *Energy,* vol. 33, pp. 817-833, May 2008.

[3] M. R. Djokic, K. M. Van Geem, G. J. Heynderickx, S. Dekeukeleire, S. Vangaever, F. Baffin-Leclerc, *et al.*, "IMPROOF: Integrated Model Guided Process Optimization of Steam Cracking Furnaces," *Sustainable Design and Manufacturing 2017,* vol. 68, pp. 589-600, 2017.

[4] C. M. Schietekat, M. W. M. van Goethem, K. M. Van Geem, and G. B. Marin, "Swirl flow tube reactor technology: An experimental and computational fluid dynamics study," *Chemical Engineering Journal,* vol. 238, pp. 56-65, Feb 2014.

[5] M. W. M. van Goethem and E. Jelsma, "Numerical and experimental study of enhanced heat transfer and pressure drop for high temperature applications," *Chemical Engineering Research & Design,* vol. 92, pp. 663-671, Apr 2014.

[6] Schmidt+Clemens. (2018, 31/01/2018). *Strengthen the heart of your plant! SCOPE Fusion HTE - the steam cracker tubing of the future*. Available: https://www.schmidt-clemens.com/products/scope/

[7] C. M. Schietekat, D. J. Van Cauwenberge, K. M. Van Geem, and G. B. Marin, "Computational Fluid Dynamics-Based Design of Finned Steam Cracking Reactors," *AIChE Journal,* vol. 60, pp. 794-808, Feb 2014.

[8] D. J. Van Cauwenberge, J. N. Dedeyne, K. M. Van Geem, G. B. Marin, and J. Floré, "Numerical and experimental evaluation of heat transfer in helically corrugated tubes," *AIChE Journal,* 2017.

[9] A. Carrillo, F. Bertola, G. Wang, and L. Zhang, "Intensified heat transfer technology–CFD analysis to explain how and why IHT increases runlength in commercial furnaces," in *AIChE and EPC 2010 Spring National Meeting, San Antonio, TX*, 2010.

572006 Control of Reactive Distillation Columns Separating Reacting Mixtures with Somewhat Unfavorable Relative Volatilities

For the separation of reacting mixtures with somewhat unfavorable relative volatilities (i.e., the reactants are the light and heaviest components and the products are the lightest and heavy ones), arranging an external recycle from the bottom to a certain stage of the reactive section onto the reactive distillation columns (RDC-ERBR) can highly improve the system steady state performance. In this article, the dynamics and control of these novel process designs separating a hypothetical reversible reaction, A + B ↔ C + D ($a_C > a_A > a_D > a_B$), were studied. With the arrangement of the external recycle, the severe asymmetry between the two product compositions and the most lightest reactant feed flow rate can be suppressed as compared with conventional reactive distillation columns, thereby presenting favorable influences to process dynamics and controllability. Three decentralized control systems with different manipulating modes of the external recycle flow rate (i.e., it is set to be a constant value, ratioed to B feed flow rate, and used to manipulate the B composition in the sensitive stage) were developed for RDC-ERBR to evaluate its influences to the process control. The results showed that the decentralized control system of the RDC-ERBR with the external recycle flow rate ratioed to B feed flow rate has the best closed loop control performance as compared to the other decentralized control systems of the RDC-ERBR and the CRDC.

567645 Intensification of Cephalexin Synthesis in Modular Microfluidic Devices Using Electric Field

Cephalexin is an important ß-lactam antibiotic that is enzymatically synthetized from a nucleophile (7-aminodeacetoxycephalosporanic acid – ADCA) and an acyl donor (phenylglycine methyl ester – PGME). Cephalexin is thermodynamically unstable and is typically produced in the kinetic regime. Prolonged exposition of cephalexin to the enzyme results in cephalexin decomposition to ADCA. Hence, good control of reaction conditions is necessary to obtain high cephalexin yield. The problems accompanying the cephalexin synthesis can be partially avoided by using microfluidic platforms.

Cephalexin synthesis can be carried out in a homogeneous arrangement with a soluble enzyme with high catalytic activity. Cephalexin extraction from the reaction mixture limits the successive hydrolysis and allows for the enzyme reuse. Aqueous two-phase systems (ATPS) represents an enzyme-friendly alternative to the extraction provided by organic solvents. In this work, we used ATPS consisting of polyethylene glycol (PEG), phosphate salts and water. We have identified such ATPS composition that is characterized by high affinity of the top PEG phase to cephalexin and the reactants and by very low affinity to the enzyme.

In the next step, we have optimized properties of the reaction mixture to attain high cephalexin concentration in the course of the enzyme reaction. We focus on the effects of temperature, the concentrations of ADCA and PGME and their molar ratio. We identified regimes that lead to ADCA conversion to cephalexin higher than 50 % within a few tens of minutes in a batch arrangement. The batch experiments gave us the estimate on the residence time of the reaction mixture in flow-through microfluidic arrangements.

Two microfluidic reactors-separators for the cephalexin synthesis have been developed. One of them relies on the slug flow arrangement of immiscible aqueous phases. The bottom phase contains the enzyme. The reactants are introduced in droplets of the top phase. The reactants are transferred through the interface into the bottom phase where cephalexin is synthetized. Cephalexin is simultaneously transferred back to the top phase where it is stabilized due to the absence of the enzyme. After phase separation, the bottom phase with the enzyme is led back to the start of the process.

The other arrangement, which is currently tested in our lab, contains a long microcapillary that serves as flow-through reactor. Both reactants and the enzyme are introduced in one phase. The output is led to a parallel flow microextractor where cephalexin and unreacted ADCA and PGME are transported to other phase. Enzyme in the original phase is led back to the microcapillary inlet. Moreover, the parallel flow microextractor is equipped by two electrode chambers. We have found that low intensity electric field applied perpendicularly to the interface allows for 100% separation of the product from the enzyme phase. Parametrical studies relevant to both arrangement will be presented.

Selected References:

Cech, J.; Hessel, V.; Pribyl, M., *Chem Eng Sci* **2017,** *169*, 97-105.
Cech, J.; Pribyl, M.; Snita, D., *Biomicrofluidics* **2013,** *7* (5), Artn 054103.
Cech, J.; Schrott, W.; Slouka, Z.; Pribyl, M.; Broz, M.; Kuncova, G.; Snita, D., *Biochem Eng J* **2012,** *67*, 194-202.
Vobecka, L.; Romanov, A.; Slouka, Z.; Hasal, P.; Pribyl, M., *New Biotechnol* **2018,** *47*, 73-79.
Pribyl, M., *Sep Purif Technol* **2019,** *221*, 311-318.

Acknowledgement: The authors thank for the support by the grant of the Czech Science Foundation [grant no. 17-09914S].

574182 3D Printed Static Mixers for Continuous Reactive Crystallization

Crystallization is an important unit operation in the pharmaceutical industry for separation and purification of intermediate chemical compounds and active pharmaceutical ingredients. The driving force for crystallization is the supersaturation of the solution, which is defined by the difference between the solute concentration and its solubility. There are various methods to create supersaturation. In anti-solvent or reactive crystallization, different process streams are mixed together to generate supersaturation. Consequently, mixing plays an important role in the performance of anti-solvent or reactive crystallization. Initially, mixing is required to obtain a homogeneous solution to obtain predictable bulk nucleation and to prevent local nucleation, which may occur due to concentration gradients[1]. Once crystals are formed, mixing is needed to maintain the crystals in suspension. Plug-flow behavior is typically desired to provide a uniform residence time for each crystal and hence deliver uniform product quality attributes such as crystal shape and size. Tubular crystallizers can provide plug-flow behavior if mixing is fast in radial direction but slow in axial direction. One of the methods to approach such mixing patterns involves static mixers. A kenics-type static mixer has previously been demonstrated to be useful for achieving plug-flow conditions for anti-solvent crystallization of ketoconazole.[2] However, other types of static mixers such has Komax, SMX, or low pressure drop (LPD) mixers have not been studied for their application as a tubular crystallizer.[3] Additionally, the mixing elements of these static mixers can be optimized for a given application.[4] Finally, the mixing requirements may change when the process material flows from the entrance to the outlet of the tubular crystallizer. In particular, in the first part of the crystallizer, rapid mixing of liquid process streams needs to be achieved, whereas large crystals need to be kept in suspension further downstream. Fouling and plugging should be avoided when the crystal mass increases. Therefore, when using a static mixer as a tubular crystallizer, the structure and dimensions of the internals may need to change for optimal operation. These requirements for customization and the absence of any moving parts call for 3D printing to be used as the preferred fabrication method of tubular crystallizers based on static mixers.

The objective of this work is to design, fabricate and characterize tubular crystallizers with different designs for static mixing for reactive crystallization. Stereolithography is identified as the most suitable 3D printing technique among the four major techniques, because it allows for fabrication of leak-proof parts and easy removal of uncured resin from the flow channel.[5] Reactive crystallization of barium sulphate is conducted in a Y-mixer and komax-type mixer fabricated using a commonly available desktop stereolithography 3D printer (Form 2, FormLabs) with clear resin. The results demonstrate practical feasibility of using 3D printing to fabricate innovative tubular crystallizers and a reduced agglomeration is observed in the komax mixer compared to a conventional Y-mixer. Subsequently, reactive crystallization of an active pharmaceutical ingredient, salicylic acid from sodium salicylate and an acid, will be investigated. COMSOL simulations show that a kenics-type static mixer with a standard design provides faster mixing compared to a Y-mixer, komax or a LPD mixer. COMSOL simulations will be used to optimize the different static mixers for the rate of mixing and to minimize zones with low liquid velocity where crystals might settle. Finally, the crystallization performance in terms of yield, crystal size distribution and degree of agglomeration will be characterized experimentally in the 3D printed tubular crystallizer. This work demonstrates a novel type of application of 3D printing in which there is a clear benefit of customization and fabrication speed.

Acknowledgement: The work described in this abstract was supported by a grant from the Research Grants Council of the Hong Kong Special Administrative Region, China, Project No. 16214418.

References

(1) Hacherl, J. M.; Paul, E. L.; Buettner, H. M. Investigation of Impinging-Jet Crystallization with a Calcium Oxalate Model System. *AIChE J.* **2003**, *49* (9), 2352–2362.

(2) Alvarez, A. J.; Myerson, A. S. Continuous Plug Flow Crystallization of Pharmaceutical Compounds. *Cryst. Growth Des.* **2010**, *10* (5), 2219–2228.

(3) Ghanem, A.; Lemenand, T.; Della Valle, D.; Peerhossaini, H. Static Mixers: Mechanisms, Applications, and Characterization Methods - A Review. *Chem. Eng. Res. Des.* **2014**, *92* (2), 205–228.

(4) Singh, M. K.; Anderson, P. D.; Meijer, H. E. H. Understanding and Optimizing the SMX Static Mixer. *Macromol. Rapid Commun.* **2009**, *30* (4–5), 362–376.

(5) Bhattacharjee, N.; Urrios, A.; Kang, S.; Folch, A. The Upcoming 3D-Printing Revolution in Microfluidics. *Lab Chip* **2016**, *16* (10), 1720–1742.

576831 Ultra-Low-Cost Continuous Manufacturing Platforms through 3D Printing and Telescoped Isolation Free Purification

Chemically resistant parts for continuous flow chemistry have been produced using the 3D printing process fused deposition modelling (FDM), from poly(etheretherketone) (PEEK). PEEK has greater chemical resistance than common FDM materials such as acrylonitrile butadiene styrene (ABS), polypropylene (PP), or even high-performance plastics like poly(etherimide) (PEI), and also has good thermal resistance and excellent mechanical strength. These printed parts have been demonstrated to be capable of withstanding high pressures, allowing the superheating of solvents for optimal reaction conditions. The ability to use FDM for these reactors allows highly customisable, cost effective flow reactors to be produced directly in the laboratory and integrated into equipment already available.

The integration of complicated internal structures, difficult to produce with powder bed fusion or resin photo-curing techniques such as selective laser melting (SLM) or stereo lithography (SL), was also possible with FDM, and allowed structures as small as 0.5 mm to be produced reliably within the reactor internal channels. These integrated elements were shown to enhance reaction mixing and the efficiency of liquid-liquid extractions, improving reaction throughput. The printed reactors were used for flow chemistry and then integrated with continuous liquid-liquid extraction followed by continuous phase separation or isolation free continuous crystallisation within an agitated bed crystalliser.

The phase separator is a compact and easily scalable setup, at a fraction of the cost of current commercial systems, requiring only simple back pressure control. The agitated bed allows continuous purification *via* crystallisation on a suspended seed bed, with the option to perform solvent swap as a final step if required. When combined with the printed reactors a customisable synthesis path results, allowing low cost integrated isolation free purification with continuous flow chemistry.

Scheme 1. (Top) 3DP reactor followed by 3DP workup stage and continuous flow phase separation in phase separator. (Bottom) 3DP reactor followed by confined suspension crystalliser

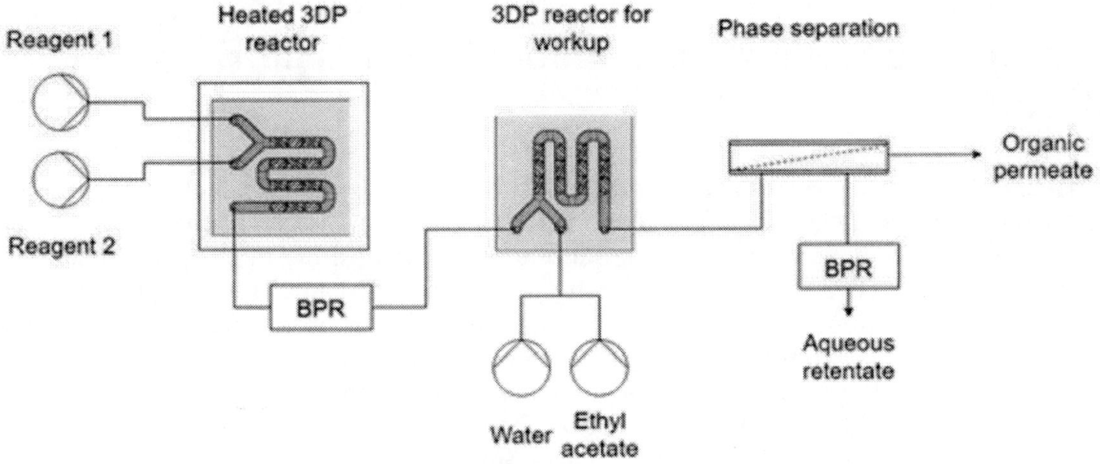
Reagent 1
Heated 3DP reactor
3DP reactor for workup
Phase separation
Organic permeate
Reagent 2
BPR
BPR
Aqueous retentate
Water
Ethyl acetate

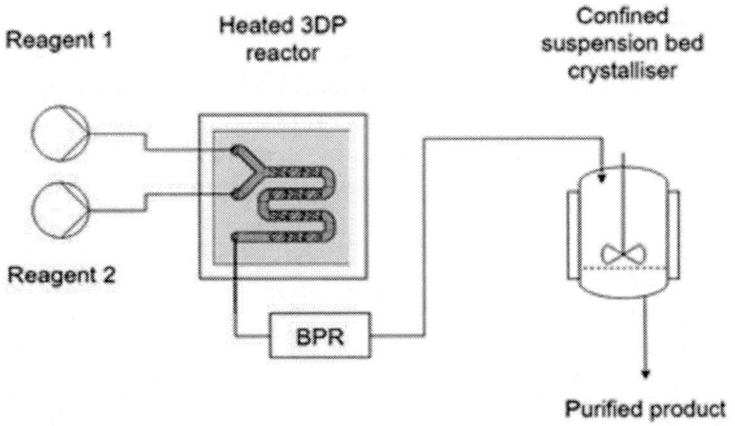
Reagent 1
Heated 3DP reactor
Confined suspension bed crystalliser
Reagent 2
BPR
Purified product

560518 Modeling of Poly(ether imide) in Material Extrusion Additive Manufacturing

Additive manufacturing (AM), is a manufacturing modality that allows the production of small scale, customized parts quickly and cheaply. Material extrusion (MatEx) AM is one of the most popular types of AM. This process uses a filament feedstock to produce a part by depositing thin roads of material in 2D layers which are then stacked in the desired geometry to create a 3D shape. However, this method of manufacturing is currently limited primarily to the rapid production of prototypes and physical models. The ultimate goal of MatEx is to produce parts for the end-use market that will experience practical use. The current state of the process prevents this due to a lack of robust mechanical properties and material versatility. The lack of versatility stems from a highly limited material catalog. MatEx is primarily commercially limited to polylactic acid (PLA) and acrylonitrile butadiene styrene (ABS). This small catalog of materials also limits the mechanical properties of the parts produced with this technique, as the mechanical properties of these two materials are much less than the common engineering thermoplastics such as polyetherimide (PEI) and polyether ether ketone (PEEK). The weakness of the produced parts also stems from a lack of z-axis, interlayer, adhesion. This problem is inherent to the process due to a low amount of time above the glass transition temperature of the material during printing, preventing the diffusion of polymer chains across the interface formed where two roads meet.

To address these issues, we propose a more thorough understanding of MatEx is required in order to more efficiently identify novel materials as well as their optimal printing conditions. Because in-situ measurements of the process are difficult, often resulting in unforeseen changes in the system that render any information gained with the measurements incorrect, we suggest utilizing multiphysics based process models to examine the effect of various process parameters on the quality of the final product. In this study, we developed multiple predictive models based on thermal transport, diffusion, and stress development. By examining the effect caused by modifying various parameters in these models, we can gain a detailed understanding of the process and how it can be modified to allow the printing of novel materials as well as identifying the optimal printing parameters for the production of parts with near bulk properties. These predictive models were then compared to empirical measurements.

577916 Characterization of Spatial Variability in Material Properties in Echogenically-Segmented Metal Components Fabricated By Additive Manufacturing

Background, Motivation, and Objective

The difference in thermal history during layer-by-layer manufacturing of metal components by selective laser melting (SLM) and related additive techniques may result in microstructural anisotropy and spatial variability in strength, density, porosity, and other material properties. The difference is particularly significant between interior and surface regions and in the build direction. While the microstructural variation may be characterized noninvasively by X-ray computed tomography, the options for nondestructive assessment of spatial heterogeneity in end-use material properties are lacking. This paper describes our progress towards the development of a novel approach to the ultrasonic characterization of spatially varying material properties and its application in additive manufacturing.

Statement of Contribution/Method

The traditional nondestructive ultrasonic evaluation gives an aggregate characterization of mechanical properties of solids and is insensitive to their spatial variabilities. This paper adapts the previously developed method for the ultrasonic measurements of segmental temperature distributions in solids [1–2] to estimate the spatial variability of mechanical properties in SLM-manufactured parts. We build upon the earlier report on assessing heterogeneity in additively manufactured aluminum parts [3] and describe our progress in designing echogenic features segmenting 3D printed parts without interfering with end-use functionality while improving the accuracy in segmental characterization.

Results

The application of the developed approach is demonstrated by characterizing segmental variability in the titanium-alloy and stainless-steel parts fabricated by SLM. Three echogenic features incorporated during the fabrication produce the train of echoes, the time-of-flight of which provides information which we use to assess the segmental variability in material properties. The ultrasonic results were found to correlate with the destructive characterization of mechanical properties. Overall, the experimental results indicate the feasibility of detecting heterogeneity of the material properties in 3D-printed parts following the developed approach.

[1] Y. Jia & M. Skliar, *Energy & Fuels*, 2016. [2] Y. Jia et al., *Ultrasonics*, 2016. [3] M. Roy, et. al., *2018 IEEE IUS*, Kobe, Japan.

575361 Lab Enablement through 3D Printing - Nitrogen Inertion Funnel

Hydrogenation reactions pose a significant fire hazard due to the use of a mix of metal catalysts, reagents and solvents. Solvent flames are especially dangerous as they are barely visible and as such a present danger for burn injuries. Most metal catalysts supported on finely divided carbon powder are highly flammable and can ignite solvents and hydrogen. The catalyst material is especially dangerous after it has been used in a hydrogenation reaction leaving the catalyst metal soaked with adsorbed hydrogen. An inertion device has been developed that reduces fire risks associated with the filtering of metal catalysts. The inertion setup is composed of a 3D printed inertion ring that clips onto commercially available filter funnels. This has been tested at two different sized filter applications to determine how much inert gas, nitrogen in this case, was required to reduce the atmospheric oxygen below 12vol% ignition level and has also been tested to see at what flow rates actual ignition is suppressed when methanol is the solvent.

3D printing was a significant factor in not only the design and prototyping of the device but also in the manufacturing of the end use devices. The final product is topologically complex. Most methods of traditional manufacturing would have added significantly to the cost and complexity. This device required non linear internal passage ways and guide vanes to direct and smooth the flow of air. To produce these devices using more traditional means would have necessitated either manufacturing in multiple parts and heat welding or combining many methods, such as sheet metal forming, machining and welding which would have significantly increased cost and limited the design in ways that could have effected the efficacy of the function.

578407 Rapid Microfluidic Production of Cell-Laden Hydrogel Microspheroids

There has been an increasing need for rapid production of microspheroidal tissues from various applications, including therapeutic cell delivery, bioreactor-based cell production, 3D disease modeling for high-throughput drug screening, and 3D bioprinting. Cells are encapsulated in hydrogels, which provide 3D scaffolds for cell proliferation and work as protections for cells during these processes. Microfluidic systems have been widely used for production of cell-laden hydrogel microspheroids. However, most current microfluidic systems require costly microfabrication facilities for device fabrication and can only produce small microspheroids (diameters < 200 μm) with low cell densities (< 10 million cells/mL) and slow production rates (polymerization time > 20s) using limited types of materials. These constraints impede their use in the downstream applications.

This study established a robust microfluidic cell encapsulation platform. Unlike other systems, our established platform uses a flexible, low cost, custom-developed molding technique for microfluidic device fabrication, which overcomes the limitations imposed by traditional photolithographic microfluidic chip fabrication. The established platform is compatible with multiple biomaterials (PEGDA, PEG-fibrinogen, and GelMA) and cell types (hiPSCs, breast cancer cells, and endothelial colony forming cells), and can produce highly uniform microspheroids (intra- and inter-batch coefficient of variance was 2% for diameter and 1% for roundness) with high cell densities (10-60 million cells/mL) and a wide range of diameters (300-1100 μm) through rapid photocrosslinking (approximate 1 s). As a result of our platform design, the encapsulated cells are evenly distributed through the microspheroids and maintain high viability and normal cellular activities in long-term culture post-encapsulation. As the major component of the platform, our microfluidic devices employ a modified T-junction design; our unique design provides control over microspheroid axial ratio and diameter and stable operation at the required high cell densities and rapid flow rates required for realizing downstream applications. We achieved high reproducibility; the resulting microspheroids are uniform both within and between batches, even when encapsulating cell clusters, which is technically challenging and is generally not feasible using microfluidic approaches.

In conclusion, this study established a robust microfluidic cell encapsulation platform fabricated with a new molding technique that enables quick device fabrication and ready testing and design iteration. Employing this platform, we rapidly encapsulated multiple types of mammalian cells in hydrogel microspheroids at much faster rates and higher cell densities than previously achieved, representing a significant advance over established systems for scale-up production of cell-laden hydrogel microspheroids.

576316 3D-Printed PLLA-Peg-PLLA Nerve Guidance Channels for Peripheral Nerve Repair

The peripheral nervous system (PNS) is vulnerable and can be easily damaged by injuries or trauma. Surgical treatment is the only effective and currently available method, with the gold standard for defects greater than 10 mm being autologous nerve grafts; however only around 40% of the 2 million US PNS patients each year regain normal function. In addition, nerve grafts have been particularly ineffective at repairing critical-size nerve defects (>3 cm). Moreover, secondary surgeries are required to harvest donor tissue, resulting in high probabilities of losing sensor and motor function at donor sites. Scaffold-based strategies where a tubular nerve guidance channel (NGC) is used to bridge the nerve defect have been promoted as a potential alternative that could avoid these additional surgeries and associated donor site morbidity. However, current NGCs lack patient-specific tunability and are only approved for small-gap (< 3cm) injuries by the U.S. Food and Drug Administration (FDA). Current research efforts are focused on creating more complex NGCs that can support regeneration of critical-size defects.

Our research seeks to use additive manufacturing technologies to create bioactive and cellular NGCs on demand for the repair of critical-size nerve defects. Recently, 3D printing has been increasingly used in research and medical therapeutics for rational, computer-aided design of biomaterial-based scaffolds with complex architecture. The NGCs should contain an outer flexible shell that seeks to mimic the mechanical properties of the surrounding biological tissue and enable diffusion of nutrients to support encapsulated cells. The use of biodegradable block copolymers with both hydrophilic and relative hydrophobic functions can provide a flexible, partially-hydrated, biocompatible and bioresorbable NGC shell. Thermoplastic elastomers (TPEs), which melt and flow above a critical temperature, turn into flexible solids upon cooling, and can be 3D-printed in a scalable fashion are reliable candidates for NGCs. This class of materials works by creating triblock copolymers where the end blocks can phase separate into hard domains from the flexible center block at lower temperatures to lock-in an elastic structure.

A-B-A type triblock copolymers of PLLA-PEG-PLLA were synthesized using varied ratios of PEG as hydrophilic groups and PLLA as hydrophobic groups. The resulting block copolymers were characterized with gel permeation chromatography (GPC), differential scanning calorimetry (DSC), and nuclear magnetic resonance (^{1}H NMR) to determine molecular weight, polymer structure, and thermal behavior. GPC results showed increasing molecular weights and decreasing polydispersity with increasing PLLA concentration in the reactor, indicating longer copolymer chains were synthesized. DSC results showed that increasing PEG block length increased the melting temperature, while increasing PLLA block length led to a decrease. Groups with melting temperatures above body temperature (37 °C) and lower melting enthalpies appropriate for 3D printing were identified. In addition, equilibrium water content, degradation rates, mechanical properties, and cell response were all evaluated and correlated to polymer structure. The results indicated that these are promising materials with high tunability that can enable patient-specific and scalable production of NGCs on-demand and on-site for the treatment of peripheral nerve injuries.

Key words: Poly (L-lactic acid); Poly (ethylene glycol); Thermoplastic Elastomers (TPEs); Peripheral nerve repair

Vascularized Multi-Tissue Platforms – Bioprinting an Organ Model *in Vitro*

Martin L Tomov[1], Andrea S Theus[1], Alexander D Cetnar[1], Holly Bauser-Heaton[1,2,3], and Vahid Serpooshan[1,2,3]

1** Coulter Department of Biomedical Engineering; Emory University and Georgia Inst. of Tech., Atlanta, GA;* ***2 *Department of Pediatrics, Emory University School of Medicine, Atlanta, GA;* ***3*** *Children's Healthcare of Atlanta, Atlanta, GA, USA*

Abstract

There is a growing need for functional *in vitro* models of tissues and organs which can be used in regenerative medicine to supplement or restore lost function and as next-generation disease modeling and drug evaluation platforms to screen candidate compounds in organ-mimic cultures. Specifically, of interest to our group are models of perfusable, vascularized tissues, as they can be applied to a range of organs and conditions that are responsible for substantial healthcare and economic burdens in the US and globally. In recent years, 3D bioprinting has emerged as a promising solution to address the need for producing consistent and reproducible tissue and organ constructs. While there have been significant advances in the field of tissue bioprinting, such as enhanced print fidelity and resolution, and better understanding of 3D culture determinants, their applications are limited by the lack of robust, guided, and reproducible vascularization and perfusion. We present here a perfusable vascularized tissue platform based on the state-of-the-art 3D printing and bioprinting that allows for the generation of complex *in vitro* multi-cellular thick tissue analogues. We created model vasculature networks using cyto-GFP endothelial cells (ECs) in various geometries and sizes, ranging from 0.5 to 2 mm, using hybrid 3D printing techniques. Our constructs, with thickness of 6 mm, were maintained under biomimetic flow conditions for 14 days. AlamarBlue assay and fluorescence microscopy showed robust increase in cell viability (metabolic activity) throughout the 14-day perfusion period. Incorporated ECs formed an endothelium on the surface of printed microchannels and initiated new capillary formation. Further, when compared to static conditions, the flow perfused constructs showed enhanced media/nutrients diffusion control, based on non-toxic dye gradient measurement. Together, our results address the need for reliable perfusion in biomimetic vascularized tissues which is a critical requirement for generation of clinically relevant thick bioprinted constructs.

575610 Bioprinting Vascularized Cancer Models

The process to develop novel therapies for cancer patients is very costly and slow due to the lack of reliable preclinical screening models. *Therefore, there is an urgent need to develop novel patient-specific cancer models.* Engineered *in vitro* tissue models utilizing patients' own cells could potentially revolutionize the way we develop treatment and diagnostic alternatives. A wide range of *in vitro* tissue models have been developed using natural and synthetic polymers. Although these *in vitro* models have been widely utilized to study tumor microenvironment (TME), majority of them are two-dimensional and don't include vasculature (crucial for tumor growth and spreading). In addition, they all display static properties, and don't mimic the dynamic changes in the TME. In this study, we present a novel 3D bioprinting approach to fabricate vascularized tumor models using hyaluronic acid based bioinks. The bioink formulation developed in this study allows stiffening behavior in the presence of cells and enzymatic degradation to enable cell migration. This platform could potentially used for semi-combinatorial drug screening, and ultimately used to develop fast and affordable cancer treatment alternatives.

577818 Chaotic Printing: Using Chaos to Fabricate Densely Packed Micro- and Nanostructure at High Resolution and Speed

Nature generates densely packed micro- and nanostructures to enable key functionalities in cells, tissues, and other materials. Current fabrication techniques, due to limitations in resolution and speed, are far less effective at creating microstructures. Yet, the development of extensive amounts of surface area per unit of volume will enable applications and manufacturing strategies not possible today. Here we describe the use of chaotic flows for rapid generation of complex, high-resolution microstructures, a technique that we call chaotic printing.

We use two classic mixing systems, the Journal Bearing (JB) Flow and the Kenics static mixer (KSM), as models to demonstrate the use of chaotic printing. In a miniaturized JB flow (miniJB), we induced deterministic chaotic flows in viscous liquids. These flows deform an "ink" (i.e., a drop of a miscible liquid, fluorescent beads, or cells) at an exponential rate to render a densely packed lamellar microstructure that is then preserved by curing or photocrosslinking. In a continuous version of chaotic printing, we created chaotic flows by coextruding two streams of alginate (two inks) through a printing head containing an online miniaturized Kenics static mixer with multiple mixing elements (or sections). In this way, we continuously 3D-printed multi-material lamellar structures with different degrees of surface area (as a function of the number of elements used) and full spatial control of the internal microstructure. The combined outlet stream was then submerged in a calcium chloride solution in order to cross-link the emerging alginate fibers and preserve the microstructure.

We show that the exponentially fast creation of fine microstructure achievable through chaotic printing exceeds the limits of resolution and speed of the currently available 3D-printing techniques. Moreover, we show that the architecture of the microstructure to be created with chaotic printing can be predicted using computational fluid dynamic (CFD) techniques.

We present different proof-of-principle applications for this technology, including the development of densely packed biocatalytic surfaces and highly complex multi-lamellar and multi-component tissue-like structures for biomedical applications

574017 3D-Bioprinting of Spatially Organized Bacterial Microcosmoi Using Chaotic Flows

Cells do not work alone but as collaborative micro-societies. Cells in human tissues are spatially organized, and these deposition patterns may have significant effects on their functionality. In the case of bacteria, spatial distribution (micro-biogeography) has also been suggested as a determinant of their behavior. Unfortunately, current microbiological techniques, which are more focused on the culture of single bacterial strains or well-mixed bacterial communities, fail to reproduce the micro-geography of polybacterial societies. Moreover, printing spatially-controlled scaffolds at high resolution is currently challenging.

Our group previously reported the development of a 3D-printing technique (chaotic bioprinting) using a Kenics static mixer to produce alginate-based and agar-based fibers exhibiting internal lamellar microstructures. In this contribution, we use chaotic bioprinting protocols to create fine-scale bacterial microcosmoi. This straightforward approach allows us to place various bacterial strains in these constructs to analyze how their spatial distribution may affect their social behavior and/or survival abilities. We demonstrate that these biological microsystems can exhibit a competitive dominance as a function of culture time, the development of hypoxic regions, and the degree of interface shared between the microcolonies.

Chaotic bioprinting enables the printing of cell-laden constructs with fine-scale deposition patterns. We envision that this technique will contribute to the development of more complex poly-bacterial microcosmoi, such as gut-microbiota models.

574032 Flow-Induced Crystallization in Materials Extrusion Additive Manufacturing

Semicrystalline polymers are often used in additive manufacturing (AM) processes due to the ease of processing in the melt state and the rapid, dramatic increase in mechanical properties during crystallization. Despite the benefits of AM, the uncertainty in the quality of the printed part makes more widespread implementation challenging. Much of the uncertainty in semicrystalline polymer AM stems from the complex relationship between processing conditions and the resulting semicrystalline microstructure. Here, we present a systematic study of printing conditions on the semicrystalline morphology of parts printed from poly(lactic acid) (PLA). The slow crystallization kinetics of PLA generate parts with low crystallinity, however a secondary annealing process at high temperatures (140 °C) generate a space-filled spherulitic texture, with smaller spherulites near the weld zone between the extruded layers. This spherulite size distribution is attributed to a higher nucleation density templated into the part by the temperature and deformation history of the printing process. We show that a combination of materials characterization and process line infrared thermography can be used in theoretical continuum modeling to predict the spatial variation in spherulite size.

570855 Computational Fluid Dynamics Simulation of the Fused Deposition Modeling Process: Effect of Temperature-Dependent Viscosity on the Shape of Extruded Strand

Fused Deposition Modeling (FDM) is a 3D printing technique based on extrusion of polymer melt. In the FDM process, polymer filaments are melted in the liquefier and then extruded layer by layer to build final products. Although the process is fast and cost-effective, the final product may have rough surface due to the deformation of the extruded strain. In this study, the effects of various process parameters on the morphology of the extruded strand have been investigated. Non-isothermal free surface flow simulation of the extrusion process was performed utilizing OpenFOAM, an open source simulation software, along with our development of a user-defined module. Simulation results for the dimensions (including thickness and width of layer) of an extruded strand have been validated by experimental results. The temperature effect on the surface roughness based on these will be also discussed.

565637 Effect of Polypropylene Melt Flow Index on the Material Extrusion Based Additive Manufacturing of Polypropylene/Hydrogenated Resin Blends

Material extrusion (MatEx) based additive manufacturing (AM) process can produce parts with intricate features that are difficult to fabricate using traditional subtractive manufacturing techniques. However, the MatEx process suffers from two major drawbacks: the mechanical properties of the parts produced using MatEx are often inferior to the bulk material properties and only a limited number of materials are currently compatible with this mode of AM. Isotactic polypropylene (PP) is a popular thermoplastic material which undergoes rapid crystallization that leads to significant trapped residual stress when processed using MatEx, resulting in poor geometric tolerance and mechanical performance.

In this work, PP having different melt flow indices (MFIs) were blended with low molecular weight hydrocarbon resins having different degrees of hydrogenation. The influence of the PP MFI on the crystallinity of the pure polymers and PP/hydrogenated resin blends as well as the melting and crystallization temperature was investigated. For example, the addition of the resins lowered the melting temperature by 7^0C while the crystallization temperature of pure PP (MFI=12g/10 min) decreased from 133.3^0C to 118.6^0C in the blends; thereby increasing the time required for the material to crystallize from the melt. Next, the blends were extruded using a single screw extruder to obtain filaments that can be used in the MatEx process. An understanding of the modifications in the crystallization process significantly impact the z-axis adhesion and residual stress state, which directly affect mechanical properties and warpage in the printed parts. Tensile bars were printed in two different orientations to analyze the mechanical performance and study the effect on part isotropy. The maximum tensile strength and modulus of the printed parts was found to decrease by 21% and 15% respectively. However, moderate post-printing annealing achieved mechanical properties that approached injection molded PP parts by improving the interlayer bonding of the parts. We can leverage these results to modify processing conditions to optimize mechanical properties of the parts generated using MatEx.

569034 Engineering Filament Design for Enhanced Printability and Mechanical Properties of Additively Manufactured Plastic Parts

3D printing enables the net shape manufacturing of objects with minimal material waste and low tooling costs, but the functionality and performance of these objects is generally limited. One of the most common 3D printing technologies and the primary consumer-user technology is fused filament fabrication (FFF), which relies on the extrusion of thermoplastic filaments through a heated nozzle. The melted filament is pushed through the nozzle by plunger action of the unheated filament. The object of interest is built up layer-by-layer as the nozzle is translated in the x-y plane. These 3D printed parts tend to suffer from poor mechanical properties due to weak interlayer interfaces that develop during the printing. There is generally an intrinsic trade-off between part accuracy and mechanical properties, where higher temperatures tend to enhance diffusion of polymer chains across the interfaces but this also tends to allow the material to flow which distorts the shape. Here we discuss routes to manipulate these trade-offs through structured core-shell filaments. Routes to produce parts with less warp (deformation) and improved mechanical properties will be discussed. Finally, we will briefly discuss how the shape of the object affects these properties and the importance of the tool path to obtain desired properties.

576383 3D Printing of Interdigitated Dielectric Elastomer Actuators

Among soft actuators, dielectric elastomer actuators exhibit excellent energy density and strain rate. However, typical fabrication methods are based on sequential planar processing, which results in planar devices with low fabrication rate. 3D printing provides the potential to rapidly fabricate versatile designs of actuators with high throughput. Interdigitated DEAs were fabricated by 3D printing vertical arrays of electrodes followed by casting a dielectric. Minimizing print defects such as end beads in the vertical electrodes is important for eliminating premature breakdown. Print defects are reduced by optimizing the ink rheology; the electrode ink is composed of a vinyl-terminated poly(ethylene glycol ethylene sulfide) oligomers with carbon black as the rheological modifier and conductive filler. The print fidelity is optimized by reducing the molecular weight of the oligomer. However, in order to achieve a low elastic modulus in the cured state, a high molecular weight between crosslinks is required. We implemented a strategy of simultaneous chain extension and crosslinking to achieve low-modulus, highly extensible elastomers using low-molecular weight precursors.

The dielectric is a chemically crosslinked oil gel with a dielectric constant of 6. The acrylate crosslinking mechanism allows chemical bonding with the electrodes, providing structural integrity in the device. The oil component in the dielectric fulfills several purposes, including reducing the uncured viscosity to improve infilling in the electrodes, reducing the tangent delta in the cured state, and providing self-healing characteristics. Self-healing occurs because oil molecules damaged by charge leakage can diffuse away from the damaged site, allowing the actuator to be actuated again.

Multinozzle 3D printing enables rapid fabrication of contractile actuators with actuation strains up to 9%. The electrodes can be fabricated at a rate of 0.73 cm^3/min; a device with an area of 7.5x3.9 cm^2 can be fabricated in ~6 min. The use of 3D printing enables versatile designs, including actuators with orthogonal pixels and prestrain-free rotational actuators.

562611 Direct Print Additive Manufacturing of Optical Fiber Interconnects for Board Level Computing Devices

Integrated photonics technology has many compelling advantages for computing and communications applications, including high speed and extremely wide bandwidth operations. Current systems are typically hybrid assemblies of packaged photonic devices where printed circuit boards often serve to route electrical signals and power, and in some cases, have runs of optical fibers. Development of fully integrated photonic systems would allow for higher transmission rates, lower power requirements, improved signal integrity and timing, less heat generation, and improved security of communication signals. Current technology has not been able to overcome how to densely route optical interconnects through tight spaces without losses due to scattering, how to easily connect optical interconnects to devices and finally how to do this in cost effective packages. Here we show that by using a new direct print additive manufacturing (DPAM) process of fused deposition modeling (FDM) of plastic and micro-dispensing of pastes and inks, along with picosecond laser processing, we can 3D print single and multi-mode optical fibers in a controlled manner such that compact, 3-dimensional optical interconnects can be printed along non-lineal paths. We have produced working optical interconnects with fibers core diameters from 77 microns to as small as 12 microns. Our results demonstrate surface roughness in 3D printed 77 micron-diameter optical fibers of less than 450 nm, with optical transmission rates above 46% for proof of concept devices. We anticipate our proof of concept devices to be a starting point in the development of more sophisticated electro-optical computing devices using this new DPAM technique.

568610 Cure-through Correction in Continuous Stereolithographic 3D Printing

Recently, continuous stereolithographic 3D printing (SL) technologies have been developed which significantly increase print speeds. By introducing a *dead zone* adjacent to the window where polymerization does not occur, continuous SL eliminates several time-consuming steps present in traditional SL and enables speeds up to 2000 $mm \cdot h^{-1}$. Two methods have been reported to generate the dead zone: (i) oxygen inhibition, and (ii) dual-wavelength exposure with complementary photoinitiation and photoinhibition.

In continuous SL, the print speed must be limited to ensure high fidelity of the printed part. Accuracy along the vertical axis is affected by the penetration depth of light into the resin. When light penetrates too deep into the resin bath, unwanted curing (known as *cure-through*, *overcure*, *the backside effect*, or *print-through error*) can occur, particularly on the backside of exterior surfaces and into internal voids. Cure-through also contributes to dose heterogeneity, introducing internal stresses which can deform the part and further reduce fidelity. Cure-through has typically been mitigated by adjusting the resin formulation to control penetration depth; however, highly absorbing resins require either higher-intensity light sources or, more commonly, slower print speeds.

Alternatively, high fidelity can be achieved without reduced print speeds by modifying slices to account for expected cure-through. This approach requires both a curing model and a slice correction algorithm. While slice correction has been developed for both external features and internal voids in non-continuous SL, existing models for continuous printing are not tailored to this purpose.

We have developed a dose-based curing model and a slice correction algorithm to precisely tune exposure levels and minimize cure-through artifacts in continuous SL. Our approach differs from previous slice-modification methods in several ways. First, we are considering a continuous system, where cure-through is a more significant and complex problem. Second, we are developing corrections using greyscale, while previous approaches were restricted to white and black. Finally, our correction uses an exact mathematical solution, while previous approaches have used heuristics and optimization.

The mathematical approach will be presented along with theoretical and corresponding experimental results from a dual-wavelength continuous SL printer. Our model allows prediction of the printed part resulting from a given set of slices and analysis of the contributions of each slice to the accumulated optical dose at each point. We have also experimentally investigated the effects of dose heterogeneity to determine optimal dose profiles. The method is demonstrated to yield marked improvement in print fidelity while maintaining high print speeds.

578293 Multiscale Modeling of Liquid-Liquid Phase Transfer Catalysis for the Simultaneous Extraction and Conversion of Hydrogen Sulfide to Value-Added Products

Phase transfer catalysis (PTC) is a general methodology applicable to a wide variety of reactions in which inorganic and organic anions, carbenes and other active species react with organic compounds. The basic principle of PTC is that the reactions take place in heterogeneous two-phase systems (organic-aqueous) with negligible mutual solubility of the phases. The organic phase consists of organic reactants (neat or in organic solvents), whereas the aqueous phase contains inorganic salts/bases. The catalyst, located in the aqueous phase, acts as a source of lipophilic cations and continuously introduces the reacting anionic species in the form of lipophilic ion pairs into the organic phase. This type of two-phase system offers numerous advantages, such as high yields and purity of the products, operational simplicity, mild reaction conditions, suitability for large-scale synthesis, and an environmentally benign nature of the reaction system.[i,ii,iii]

However, the downside of PTC is the need to quantify exacting reaction conditions and parameters that are difficult to uncover and, in some cases, are counterintuitive.[ii] In a 1998 comprehensive review of the field from an engineering perspective,[iv] the authors lamented the lack of published theoretical process modeling for scale-up and concluded that this has hurt the commercialization potential of PTC. Although the next two decades have seen some progress in terms of mathematical modeling of PTC systems,[v,vi,vii] availability of accurate thermodynamic parameters still proves a major limitation as the chemical domain in the group contribution methods, such as UNIFAC, is inherently limited to the portion of the chemical design space for which every binary interaction parameter is available.

In this respect, COSMO-based models, such as COSMO-RS,[viii,ix] are valuable alternatives for describing liquid-phase thermodynamics with the major advantage of not requiring any binary interaction parameters to calculate chemical potentials and, consequently, free energies of solvation. Furthermore, COSMO-RS allows for easy integration of quantum chemical calculations into a process modeling framework, greatly expanding the envelope of chemical species that can be modeled at a high level of accuracy.

The primary objective of this work is to develop a rigorous multiscale mathematical model to demonstrate the application of PTC to hydrogen sulfide recovery and conversion from an aqueous alkanolamine solution to value-added organic products to improve process economics and sustainability. Desulfurization of sour natural gas has been an increasing concern for the energy industry due to the high economic and environmental cost of the existing technologies. Application of PTC to this problem can effectively overcome these issues by providing a waste-to-value pathway toward resource recovery and circular economy. With Python as the high-level interface, the solution algorithm uses varying tools for different scales of time and space to model and predict the behavior of the system: Turbomole at quantum mechanical/continuum solvation level, COSMOtherm at solvent/mixture level, and Pyomo at process/systems level. A dynamic-programming approach is employed to solve the resulting nonlinear optimization problem using the IPOPT solver.

By combining these tools in a novel integrated scheme, we can identify the extent of possible reaction pathways leading to the design of intensified unit operations in an optimization-based integrated product-process design framework. The proposed methodology can be furthered used as a basic platform to simulate similar systems. Herein, the multiscale model described above is applied to the identification of catalyst/product(s), optimization of the organic solvent and design of the intensification process, turning the hazardous by-product of natural gas processing into a value-added product.

[i] Mieczyslaw Makosza and Michal Fedorynski. Phase Transfer Catalysis. Catalysis Reviews-science and

Engineering, 45:321-367, 2003.

[ii] Domenico Albanese. Liquid-Liquid Phase Transfer Catalysis: Basic Principles and Synthetic Applications. Catal. Rev. - Sci. Eng., 45(3-4):369-395, 2003.

[iii] Seiji Shirakawa and Keiji Maruoka. Recent developments in asymmetric phase-transfer reactions. Angew. Chemie - Int. Ed., 52(16):4312-4348, 2013.

[iv] Sanjeev D. Naik and L. K. Doraiswamy. Phase transfer catalysis: Chemistry and engineering. AIChE J., 44(3):612-646, 1998.

[v] Amata Anantpinijwatna, Mauricio Sales-Cruz, Sun Hyung Kim, John P. O'Connell, and Rafiqul Gani. A systematic modelling framework for phase transfer catalyst systems. Chem. Eng. Res. Des., 115:407-422, 2016.

[vi] Chiara Piccolo, Andrew Shaw, George Hodges, Patrick M. Piccione, John P. O'Connell, and Rafiqul Gani. A framework for the design of reacting systems with phase transfer catalysis. Comput. Aided Chem. Eng., 30:757-761, 2012.

[vii] Justinus A.B. Satrio and L.K. Doraiswamy. Phase-transfer catalysis: a new rigorous mechanistic model for liquid-liquid systems. Chem. Eng. Sci., 57(8):1355-1377, 2002.

[viii] Andreas Klamt. Conductor-like screening model for real solvents: A new approach to the quantitative calculation of solvation phenomena. Journal of Physical Chemistry A. 99(7): 2224-2235, 1995.

[ix] Andreas Klamt, Volker Jonas, Thorsten Bürger, & John C.W. Lohrenz. Refinement and parametrization of COSMO-RS. Journal of Physical Chemistry A. 102(26):5074–5085, 1998.

564195 Optimization of Syngas to Olefin (STO) Reactors

Optimization of Syngas to Olefin (STO) Reactors

Can Ekici[1], Lorenz T. Biegler[1], Joseph F. DeWilde[2], Dylan Kipp[2], Paul M. Witt[2]

1: Carnegie Mellon University, 2: The Dow Chemical Company

Reactor designs with mixed catalysts play an important role transforming a multiple reactor system to single-shot reactors. In addition to savings in capital and ease of implementation, single-shot reactors are useful to break equilibrium limitations, therefore increasing the yield and selectivity of desired product as shown in [1, 2]. However, the nonlinear and highly exothermic nature of mixed-catalyst systems makes it difficult for commercial process simulation and optimization tools to optimize these systems. This talk describes the development and application of optimization strategies for mixed-catalyst, single-shot reactors for syngas to olefin (STO) processes.

Finding the optimal catalyst distribution is challenging and requires advanced solution strategies for singular optimal control problems, which are poorly conditioned and often lead to flat response surfaces. The graded bed approach is applicable to these problems for preliminary results, but more sophisticated solution approaches are needed in order to find the exact optimal profiles. To determine these profiles, we develop a partial moving finite element approach. Starting from an equally spaced coarse grid and calculating the errors on non-collocation points, fixed elements are introduced at points where error constraints are violated. and moving elements to determine optimal breakpoint locations. Finally, the spikes are detected and eliminated, in order to ensure that optimality conditions are satisfied. Embedded within a nonlinear optimization strategy, this approach decreases the complexity and solution times significantly compared to previous singular optimal control optimization strategies [3].

The resulting optimal control strategy is applied to a comprehensive STO reaction mechanism with detailed rate expressions that extends the mixed-catalyst STO reaction mechanism in [4]. Moreover, a parallel experimental study to reaction mechanism is currently underway for better understanding and representation of the phenomena.

Finally, the overall goal of this RAPID project is to find the optimal catalyst mixing ratio along the bed for the mixed catalyst STO reactor, respecting safety and quality constraints. The two optimization approaches (graded bed and partial moving grid) will be applied to an evolving to STO reaction model, with updates due to possible changes in reaction mechanism as parallel experimental studies continue.

[1] D.L.S. Nieskens, A. Ciftci, P.E. Groenendijk, M.F. Wielemaker, A. Malek, I&EC Research, 56, 2722-2732 (2017)

[2] S. K. Mazidi, M. T. Sadeghi, M. A. Marvast, Chem. Eng. Technol. 2013, 36, No. 1, 62–72

[3] W-F Chen, L. T. Biegler, "A Simultaneous Approach for Singular Optimal Control Based on Partial Moving Grid," AIChE Journal, to appear (2019)

[4] Alexey V. Kirilin, Joseph F. Dewilde, Vera Santos, Adam Chojecki, Kinga Scieranka, and Andrzej Malek, Ind. Eng. Chem. Res. 2017, 56, 13392-13401

569006 Sustainable Process Intensification Using Building Blocks

Growing demands for chemical products and increasing competition in the global market pose new challenges on the chemical process industry. Chemical processes need to produce more with less raw materials and utilities while generating less waste and reduced emissions. New projects are also required to consider sustainability of the process besides the economics throughout the project life-time. Process intensification (PI) provides means for more sustainable, cost effective, smaller and safer designs compared to their conventional counterparts [1]. Several systematic process intensification methods have been developed in the past that considers the trade-offs between sustainability and economics in process design [2-5]. These trade-offs can be illustrated through optimal Pareto fronts which, for given design alternatives, often describe the limits of our design achievements.

In this work, we ask whether systematic process intensification can provide novel solutions that are not bounded with these trade-offs and lead simultaneously more sustainable and economic designs. To this end, our recently proposed building block-based approach [6-9] provides several unique advantages towards incorporating both economic and sustainability objectives in process synthesis and intensification. Unlike traditional synthesis approaches that require superstructures with pre-postulated unit operation alternatives, our building block-based superstructure relies on physicochemical phenomena to automatically generate optimal intensified flowsheets. It is formed by building blocks positioned on a two-dimensional grid. A mixed-integer nonlinear optimization (MINLP)-based model is used to describe the superstructure. With this, novel intensified flowsheets can be generated through minimizing/maximizing several different objectives, e.g. economics, waste generation, utility consumption, etc. Specifically, we will introduce sustainability considerations through multi-objective optimization and search for pareto-optimal intensified flowsheet variants. This provides a unique approach for process design and intensification in which the effects of intensification on the process economics as well as on the sustainability can be observed. We will demonstrate the benefits of this design approach through several case studies with special focus on energy-intensive separations and reactive separation processes.

References:

1. Stankiewicz, A. I., Moulijn, J. A. (2000). Process Intensification: Transforming Chemical Engineering. Chemical Engineering Progress 1, 22–34.
2. Demirel, S. E., Li, J., & Hasan, M. M. F. (2019). Systematic process intensification. *Current Opinion in Chemical Engineering*. https://doi.org/10.1016/j.coche.2018.12.001
3. Tian, Y., Demirel, S. E., Hasan, M. M. F., Pistikopoulos, E. N. (2018). An Overview of Process Systems Engineering Approaches for Process Intensification: State of the Art. *Chemical Engineering and Processing: Process Intensification*, 133, 160-210.
4. Tula, A. K., Babi, D. K., Bottlaender, J., Eden, M. R., Gani, R. (2017). A computer-aided software-tool for sustainable process synthesis-intensification. *Computers & Chemical Engineering*, *105*, 74-95.
5. El-Halwagi, M. M. (2017). A return on investment metric for incorporating sustainability in process integration and improvement projects. *Clean Technologies and Environmental Policy*, *19*(2), 611-617.
6. Demirel, S. E., Li, J., and Hasan, M. M. F. (2017). Systematic Process Intensification using Building Blocks, *Computers and Chemical Engineering*, 105, 2-38.
7. Li, J., Demirel, S. E., & Hasan, M. M. F. (2018). Process integration using block superstructure. *Industrial & Engineering Chemistry Research*, *57*(12), 4377-4398.
8. Li, J., Demirel, S.E. and Hasan, M.M.F. (2018). Process synthesis using block superstructure with automated flowsheet generation and optimization. *AIChE Journal*, 64(8), 3082-3100.
9. Demirel, S. E., Li, J., and Hasan, M. M. F. (2019). A General Framework for Process Synthesis, Integration, and Intensification. *Industrial & Engineering Chemistry* DOI: 10.1021/acs.iecr.8b05961.

565754 Batch to Continuous (B2C) Reactor Intensification through a First-Principles-Based Optimization Framework

The conversion of existing batch processes to novel, intensified continuous flow processes has been the subject of a significant research effort in the last two decades. Continuous flow processes generally promise a smaller physical footprint, while offering many technical advantages such as better heat and material management, established process control principles, and simpler (and faster) scale-up from the bench to the industrial scale [1].

Among the main challenges involved in making the B2C transition for chemical reactors is determining whether this transition is feasible. For example, chemical processes involving slow reactions are often conducted in batch reactors, since an equivalent continuous flow reactor (CFR) at the same operating conditions would be unreasonably long. Nevertheless, the B2C switch under these circumstances is not necessarily impossible, if changes in operating conditions (e.g., temperature, temperature profile) are considered. Therefore, assessing the feasibility of the switch must take the specific constraints on the different equipment involved in the two processing modes into account.

For this reason, feasibility studies that focus on assessing the possibility of making the B2C reactor switch often involve testing a particular reaction (experimentally or computationally) in batch and CFRs independently, then comparing their performances. General approaches (applicable to any chemical reaction scheme) to assess feasibility are scarce, and are mostly based on heuristics (see, for example, [2]).

Motivated by the above, in this work, we develop a general, optimization-based framework that utilizes existing batch reactor information to assess B2C feasibility. We begin by assuming that a reference batch process exists, and that it is optimal with respect to some defined metric. Then, starting with first-principle models, we define dimensionless performance metrics that are applicable for the two reactor types, and show the equivalence between the evolution (trajectory) of these metrics in time (for the batch reactor) and space (for the continuous flow reactor). This then allows us to use concepts from optimal control to assess the feasibility of designing CFRs whose performance (defined in terms of said "trajectories") are equivalent to the reference batch system. Multiple case studies are presented to illustrate the theoretical developments.

References

[1] Hessel, V. Novel Process Windows – Gate to Maximizing Process Intensification via Flow Chemistry, Chemical Engineering Technology, 32, 1655-1681, 2009

[2] Plouffe, P, Macchi, A, Roberge, D. From Batch to Continuous Chemical Synthesis – A Toolbox Approach, Organic Process Research & Development, 18, 1286-1294,2014

574405 Development of Process Operability Algorithms for Modularization and Intensification of Energy Systems

A few decades ago, the measure of process output controllability motivated the conception of the process operability approach along with its operability index (OI) [1] that is calculated based on input-output mappings. Recent advances on the operability framework [2-4] have explored the extension of the operability inputs, traditionally characterized by potential manipulated variables, to include design variables (physical dimensions, material properties, etc.). The extended approaches verified the capability of selected designs to achieve desired production and efficiency goals. In particular, process intensification (PI) and system modularization (SM) targets were considered for improved process efficiency, and an optimization-based framework was formulated to find an optimal modular design. Nevertheless, the performed calculations were restricted to nominal process conditions and thus no assessment of system operation using the OI was performed, which corresponds to a gap in the SM and PI literature.

To fill this gap, in this presentation, a newly developed multilayer operability framework [5] is introduced and further developed. This framework accounts for process operations by employing the measure of OI to rank competing modular designs. The framework is further developed to consider the incorporation of key operability aspects with the aim of addressing applications of different levels of complexity. Specifically, the operability input-output mapping is augmented to accommodate both design and operational variables. To obtain a modular region containing candidate designs for SM and PI, a mixed-integer linear programming algorithm is formulated based on a multimodel representation of the energy system in focus. For the multimodel representation, computational geometry tools such as Delaunay triangulation and tessellation are applied to decompose the original nonlinear mapping into a set of paired polytopes. Some of the advantages of this system representation correspond to a straightforward model inversion and fast computation of the hypervolumes for quantification of the OI.

Several input-output mapping strategies will be examined with a focus on handling increases in system dimensionality. The inclusion of process disturbances and the incorporation of flexibility and controllability will also be addressed to enhance the operability studies towards SM and PI. To show versatility of the developed framework, a variety of applications will be considered including the direct methane aromatization conversion to hydrogen and benzene, the classical shower problem in process operability, and a power plant cycling application for power generation with penetration of renewable energy sources. The resulting operability algorithms and applications will be used for construction of an open-source operability toolbox in MATLAB.

References

1. Vinson, D. R.; Georgakis, C. New Measure of Process Output Controllability. *Process Control* **2000**, 10 (2–3), 185–194.

1. Gazzaneo, V.; Carrasco, J. C.; Lima, F. V. An MILP-Based Operability Approach for Process Intensification and Design of Modular Energy Systems. *Comput.-Aided Chem. Eng.* **2018**, 44, 2371–2376.

3. Carrasco, J. C.; Lima, F. V. Novel Operability-Based Approach for Process Design and Intensification: Application to a Membrane Reactor for Direct Methane Aromatization. *AIChE J.* **2017**, 63 (3), 975–983.
4. Carrasco, J. C.; Lima, F. V. Bilevel and Parallel Programing-Based Operability Approaches for Process Intensification and Modularity. *AIChE J.* **2018**, 64 (8), 3042–3054.
5. Gazzaneo, V.; Lima, F. V. Multilayer Operability Framework for Process Design, Intensification, and Modularization of Nonlinear Energy Systems. *Eng. Chem. Res.* **2019**, DOI: 10.1021/acs.iecr.8b05482.

569744 A Modular Approach to Process Integration and Intensification

To survive the competitive global market concerned with environmental risk and resource shortage, there is a need for the chemical/energy industry to develop innovative and sustainable processes with substantially improved cost performances, increased energy efficiency, and reduced waste generation [1-3]. Process integration and intensification from systems perspective can contribute to this challenge by providing systematic approaches for the design of integrated production systems ranging from individual processes to total sites to enhance resource utilization [4-5], as well as for the discovery of novel intensified designs by synergizing multifunctional phenomena at different time and spatial scales to boost process improvements [3,6-7]. However, key open questions remain for these synthesis approaches on: (i) what are the driving forces to move towards integrated or intensified schemes, (ii) how to systematically derive process solutions which fully exploit integration and intensification opportunities, and (iii) how to explore the combinatorial design space in a computationally efficient way.

In this work, we propose a systematic approach for process synthesis, integration, and intensification based on recent extensions of the Generalized Modular Representation Framework [8-11]. Herein, the chemical processes are represented as aggregated multifunctional mass/heat exchange modules with which intensification possibilities can be discovered without a pre-postulation of equipment or flowsheet configurations. Driving force constraints, derived from total Gibbs free energy change, are employed to characterize mass/heat transfer feasibility by exploiting the general thermodynamic space, and thus result in a more compact modular representation of the chemical systems. Mass and/or heat integration are simultaneously addressed in the superstructure representation of the modular network, without pre-postulation of steam properties as rich/lean streams or hot/cold streams. The resulting synthesis problem is formulated as a mixed integer nonlinear programming optimization problem (MINLP), where both process cost performances and environmental regulations can be accounted for. Three case studies are presented to showcase the proposed approach for process integration and intensification, namely: (i) heat-integration process, (ii) extractive separation process with heat & mass integration, (ii) reactive separation process.

Reference

1. Bielenberg, J., & Palou-Rivera, I. (2019). The RAPID Manufacturing Institute – Reenergizing US Efforts in Process Intensification and Modular Chemical Processing. *Chemical Engineering and Processing-Process Intensification*.
2. Tian, Y., Demirel, S. E., Hasan, M. M. F., & Pistikopoulos, E. N (2018). An Overview of Process Systems Engineering Approaches for Process Intensification: State of the Art. *Chemical Engineering and Processing: Process Intensification*, 133, 160-210.
3. Tula, A. K., Babi, D. K., Bottlaender, J., Eden, M. R., & Gani, R. (2017). A computer-aided software-tool for sustainable process synthesis-intensification. *Computers & Chemical Engineering, 105*, 74-95.
4. Li, J., Demirel, S. E., & Hasan, M. F. (2018). Process integration using block superstructure. *Industrial & Engineering Chemistry Research*, *57*(12), 4377-4398.
5. Papalexandri, K. P., Pistikopoulos, E. N., & Floudas, A. (1994). Mass-exchange networks for waste minimization-a simultaneous approach. *Chemical engineering research & design*, *72*(3), 279-294.
6. Pichardo, P., & Manousiouthakis, V. I. (2017). Infinite DimEnsionAl State-space as a systematic process intensification tool: Energetic intensification of hydrogen production. *Chemical Engineering Research and Design*, *120*, 372-395.
7. Demirel, S. E., Li, J., & Hasan, M. F. (2017). Systematic process intensification using building blocks. *Computers & Chemical Engineering, 105*, 2-38.
8. Papalexandri, K. P., & Pistikopoulos, E. N. (1996). Generalized modular representation framework for process synthesis. *AIChE Journal*, *42*(4), 1010-1032.
9. Ismail, S. R., Proios, P., & Pistikopoulos, E. N. (2001). Modular synthesis framework for combined separation/reaction systems. *AIChE Journal*, *47*(3), 629-649.
10. Proios, P., Goula, N. F., & Pistikopoulos, E. N. (2005). Generalized modular framework for the

synthesis of heat integrated distillation column sequences. *Chemical engineering science*, *60*(17), 4678-4701.

11. Tian, Y., & Pistikopoulos, E. N. (2019). Synthesis of Operable Process Intensification Systems—Steady-State Design with Safety and Operability Considerations. *Industrial & Engineering Chemistry Research*. **DOI:** 10.1021/acs.iecr.8b04389.

577804 Mathematical Optimization and Process Intensification of Diafiltration Membrane Systems

Mathematical Optimization and Process Intensification of Diafiltration Membrane Systems

Elvis A. Eugene, William A. Phillip, and Alexander W. Dowling

Chemical and Biomolecular Engineering | University of Notre Dame

Membrane systems are superior to current separations processes because of lower footprints (compactness) and lower mass transfer limitations (rapidity) and are therefore an exciting field of research for sustainable technology [1]. However, there exists a critical need to systematically identify the most promising applications for novel membrane materials and address the gaps in scientific knowledge that most inhibit their translation to scalable technologies. Diafiltration, a continuous operating strategy for staged membrane cascades, is one such technology, but barring a few experimental lab scale demonstrations, has not been rigorously modeled and analyzed [2]. We present a rigorous superstructure optimization framework for a continuous diafiltration cascade design and intensification. In a battery recycling case study, we show how staging in diafiltration cascades could enable novel membrane separations of lithium and cobalt [3]. Through sensitivity analysis with the optimization model, we identify membrane materials property targets. We conclude with ongoing nonlinear parameter estimation work to assess different transport mechanisms and interfacial phenomena from dynamic diafiltration experiments.

In diafiltration, dialysate is strategically added to the feed channel of a membrane module to offset concentration polarization effects and achieve high purity and recovery of products [4]-[6]. We present a novel modeling and superstructure optimization framework to 1) elucidate optimal multi-stage process configurations with complex recycle strategies and 2) systematically identify property targets for membrane materials. We present strategies to study several configurations of a diafiltration process network (e.g., counter-current, co-current) using superstructure optimization. Our nonlinear optimization model treats flow and concentration of all streams in the network as decision variables. We then use the epsilon-constraint method to characterize the Pareto trade-off set between purity and recovery objectives. Interestingly, we find distinct regions of the Pareto set with similar recycle strategies but each with unique flow and concentration profiles. Using an illustrative case study, we show how novel 3 or 4-stage cascade design recycle strategies with modest improvements in membrane materials could enable new lithium and cobalt separations technologies for battery recycling [3]. We also compare our proposed diafiltration technology with classic process intensification technologies such as distillation and thereby accentuate the potential for disruptive membrane-based technologies of the future.

Guided by these membrane material performance targets, we conclude with some brief remarks about our ongoing work to elucidate molecular design rules for chemically selective copolymer membranes. These membranes use chemical factors to selectively target similarly sized solutes for separation from a feed solution. Batch diafiltration membrane characterization experiments are inherently dynamic, with membrane interfaces exposed to a large range of solution compositions. We propose a dynamic model and parameter estimation approach to quantitatively compare different membrane-solution thermodynamic models and transport mechanisms. In particular, we are interested leveraging optimal design of experiments paradigms to systematically design dynamic diafiltration experiments which can maximize information gain and better discriminate between rival models, i.e., proposed transport mechanisms [7]-[8]. We wrap up with some remarks on future opportunities for uncertainty quantification and systems engineering to guide material discovery and ultimately realize our vision of integrated engineering frameworks that span molecular to systems scales.

References:

[1] Qu, S., Dilenschneider, T., & Phillip, W. A. Preparation of Chemically-Tailored Copolymer Membranes with Tunable Ion Transport Properties. *ACS Applied Materials and Interfaces* **2015,** *7,* 19746 – 19754.

[2] Nambiar, A. M. K., Li, Y., & Zydney, A. L. Countercurrent staged diafiltration for formulation of high value proteins. *Biotechnology and Bioengineering* **2018,** *115(1),* 139–144.

[3] Eugene, E. A., Phillip, W. A., & Dowling, A. W., Material Property Goals to Enable Continuous Diafiltration Membrane Cascades for Lithium Ion Battery Recycling, *Submitted to FOCAPD 2019.*

[4] Cheryan, M. *Ultrafiltration and Microfiltration Handbook,* **1998.**

[5] Mulder, M. *Basic Principles of Membrane Technology,* **1998.**

[6] Strathmann, H. Selective removal of heavy metal ions from aqueous solutions by diafiltration of macromolecular complexes. *Separation Science and Technology* **1980,** *15(4),* 1135–1152.

[7] Franceschini, G., & Macchietto, S. Model-based design of experiments for parameter precision: State of the art. *Chemical Engineering Science* **2008,** *63*(19), 4846–4872.

[8] Laínez-Aguirre, J. M., Mockus, L., & Reklaitis, G. V. A stochastic programming approach for the Bayesian experimental design of nonlinear systems. *Computers and Chemical Engineering* **2015,** *72*, 312–324.

583942 More Sustainable Winemaking

The wine industry is facing urgent challenges to minimize energy and water usage, wine loss, waste generation while working within the constraints of labor availability. The industry has grown by often scaling up batch processes without introduction of benefits enabled by continuous or semi-continuous processes. Two such processes are those for stabilizing wines from precipitating potassium bitartrate and developing protein hazes. The traditional in-tank approaches result in excess usage of water and energy, loss of wine, and generation of solid waste. Two liquid-solid separations processes are being adapted to benefit the wine industry: Fluidized Bed Crystallization for removing excess potassium bitartrate and Ion-Exchange for removing excess proteins. These approaches are based upon similar chemical and physical interactions to those of traditional in-tank treatments. Our results, which build on prior work, continue to demonstrate their efficacy under a variety of wine chemistries and process conditions and aim to provide the industry with another alternative to existing treatments which can minimize energy and water usage, wine loss and waste generation.

553054 Evaluation of Ultrasound By Direct Contact on Convective Drying of Banana (*Musa paradisiaca*) Slices Using Principal Component Analysis and Response Surface Methodology

Conventional convective drying is one of the latest techniques used to reduce the amount of water in foods, however, due to the long times of exposure to high temperatures, these techniques can reduce the sensory quality, in addition to generating high costs associated with consumption of energy. An alternative to improve efficiency in these processes is to use innovative tools such ultrasound (US) application by using direct contact during drying to improve quality and keep nutritional content of the product with reduction in time, cost and temperatures of operation.

In this study, a Box-Behnken experimental design was used to study the influence of three independent variables: US time-frequency (t), US power level (Pot), and air temperature (T) on the drying process time, apparent density, total polyphenol content and color difference as responses of the banana slices. Seventeen experiments were performed which involved 12 three-level experimental points and 5 repetitions in the central point. Fisher's statistical testing was performed for the analysis of variance (ANOVA) for quadratic regression equations. Results of each response variable were analyzed according to the Response Surface Methodology, analyzing tendencies and regions of operation with contour graphs. In addition, models were proposed for each response variable, which included the input factors of the experimental design proposed. These models were reduced in order to eliminate non-significant terms and were carried out validation of the statistical models proposed. After performing the reduction of non-significant terms, the ANOVA analysis showed a good fit for most of the response variables. PCA was performed to evaluate effect of ultrasound on quality characteristic of banana-dried slices. The PCA analysis showed that the drying time and the apparent density of the samples was strongly related to the temperature of the process. Moreover, power level of the ultrasound influenced the color variation of the dry sample respecting fresh fruit. Finally, US time-frequency affected total polyphenol content of dried samples. Analysis of both tools like PCA and RSM allowed knowing the effect of the independent variables on the drying operation and quality parameters on banana dried slices.

561020 Energy Usage in Food Manufacture

The food industry, the largest manufacture sector globally and also one of the major contributors to GHG emissions, is facing significant environmental challenges that are building up pressure on food security. As the global population keeps growing, water, food and energy demand are expected to increase too - up to 50 % by 2050. A starting point to reduce such demand is to identify energy consumption hot spots in food manufacture operations. In this context, this work provides a comprehensive and updated database for energy processing usage – within the UK and globally. Specific energy consumption (SEC) data for products, processes and food distribution were collected from 44 literature sources based on product-based energy intensity metrics. To account for changes in technologies, processing and fuel efficiencies, and structural changes within sectors data was sorted in three groups: prior to 2000, 2000 - 2004, and 2005 – 2015.

Findings show that thermal processes consume most of the total processing energy – generated extensively from fossil fuels (e.g. > 65% average of the dairy sector), which shows potential for alternative technologies to be implemented. Emerging technologies are typically electric powered, so the energy could be supplied from renewable resources. Data also revealed that the most energy intensive food products are powders (instant coffee and milk powder), fried goods (French fries and crisps) and bread, all them products that involve phase-change in their manufacturing, such freeze-drying or drying. The ways in which many food processes are carried out require first the addition of water to the product followed by its removal (either by evaporation or sublimation), revealing water flow streams that can be optimised and recycled. In the meat and dairy processing sectors, energy and water use have increased due to a rise in hygienic standards and cleaning requirements, which represents an opportunity for hygienic design developments. The challenge here is to introduce energy and water-usage saving policies into the manufacturing protocols without either affecting the quality of the foodstuff or making it less safe. Developing food formulation with lower moisture contents could be a solution.

Energy use in the UK follows the same trends, although a lack of studies that assessed food products playing a central role to the UK – e.g. ready to eat meals, sandwiches – was also identified. An analysis of the relationship between energy density (calories intake), weekly consumption and processing energy use in the UK revealed that the some of the more demanded processed, nutrient-dense foods (e.g. cakes, cookies) require also higher amounts of energy during processing.

In terms of food distribution, a shift towards distributed supply systems – as opposed to centralised ones – and to local production has been identified. For example, more than 98% of all foods within the UK are transported by road, with increasing distances travelled in recent years. However, the environmental benefits of these changes are not always clear, which reveals the importance of global environmental assessment tools (such LCA, carbon and water footprint) that take the whole food chain into consideration.

Overall, this work reveals a series of needs and points towards current trends in food manufacturing that can help both stakeholders and policy makers to allocate resources more effectively and design new food manufacturing protocols that increase sustainability and security of food chains.

583939 Sustainability of Nano-Packaging Use for Food Shelf Life Extension: Life Cycle Assessment Perspective

Food waste is a severe economic and environmental problem in industrialized countries. Nano-packaging has recently become a promising technology for food preservation with increased shelf life, and hence has the potential to reduce food waste. However, the production of nanomaterials can generate a variety of environmental burdens, including the use of energy, water, and other resources, and the associated emissions. In this study, a cradle-to-grave life cycle assessment was conducted to evaluate the net carbon footprint of four nano-packaging systems containing selected food products, considering the direct impact of nanomaterial production and the indirect impact caused by food waste reduction. With the functional unit defined as 1 kg of food product and the required amount of nano-packaging materials, the global warming potential (GWP) associated with the nano-packaging system was highly affected by the concentration of incorporated nanomaterials and the type of packaged food. Furthermore, the packaging-food ratio played an important role in the relative contributions of nanomaterial, polymer, and food to the GWP of a nano-packaging system. A consumer behavior study was also performed to indirectly quantify the amount of food waste reduced resulting from the shelf life extension of food products. The results showed that the GWP decreased due to food waste reduction was significantly higher than the additional GWP produced by nanomaterials, indicating a net beneficial impact of the use of nano-packaging on the carbon footprint of food preservation. This proof-of-concept study presents a novel methodological framework to evaluate the environmental trade-offs of nano-packaging for food preservation, which can help the public better understand the net carbon footprint of nano-packaging from a life cycle perspective. The results of this study are expected to provide the groundwork for food manufacturers to make more informative decisions on the applications of this emerging technology in food industry.

583940 Water Sustainability: The Coca-Cola Company's Story - From Risks to Opportunity

As early as 2002 The Coca-Cola Company faced risks to its business based on water sustainability. In 2003 the Company examined various types of water risks, conducted a risk survey of manufacturing sites, and determined its risk profile and a series of mitigation strategies. Subsequent water risk assessments in 2008 and 2015 have formed the basis for additional mitigation strategies and helped establish the path forward for the Company.

572376 Building Momentum for Sustainable Behaviors in Developing Regions Using Locally Managed Decentralized Circular Economy Principles

Despite the current threat from climate change, plastic collecting in the world's oceans, and the steady loss of biodiversity, the world continually fails to take action with regard to our rapidly changing ecosystem. Unfortunately, waiting on governments to act is no longer a viable option. Rapid change is needed and the pace of diplomacy is simply too slow. Democratic governments are reactionary and taking action to solve future problems is not a priority, even as the future draws ever closer. Change is in the hands of individuals and it our decisions and behaviors that will influence the future of our planet and our ability to inhabit it. Therefore, building momentum for sustainable behavior must begin with individuals. The neoliberal approach to environmental protection posits that individuals are motivated by rational self-interest and that economic incentives are necessary to achieve environmental goals. However, recent research suggests that monetary gain alone actually negatively impacts behavior, and often neglects the rural poor. As a result, models for projects designed to benefit the environment need more than just a monetary incentive, they must incorporate all three pillars of sustainability: environment, economy and society. One approach for building momentum for sustainable behavior with regard to municipal solid waste management, particularly in the developing world, is by implementing Locally Managed Decentralized Circular Economy (LMDCE) principles. This contribution will describe the role behavioral economics plays in the choices made by producers and consumers. The results of a case study on applying LMDCE principles in Kampala, Uganda to manage waste plastic accumulation by conversion to fuel oil will be presented.

575678 Supply Chain Optimization of a Locally-Managed Decentralized Solution for Conversion of Waste Plastic to Fuel in Developing Regions

Plastic waste accumulation is a global threat. Due to the lack of sufficient capital, infrastructure and knowledge pertaining to hazards associated with waste accumulation, plastic waste management is a serious issue in rural regions of developing countries. This research proposes the implementation of a locally-managed decentralized circular economy (LMDCE) for reusing and recycling waste plastic at its source of generation. A feasible LMDCE solution is conversion of waste plastic to fuel oil via thermal decomposition, utilizing appropriate technology.

This research presents an initial supply chain optimization case study of a small-scale, low-cost, LMDCE based waste plastic to fuel oil process – designed by the University of Kentucky – in Kampala, Uganda. Utilizing MATLAB computational tool, the optimum number, size, and locations of the waste plastic to fuel technology will be determined. The role of small entrepreneurs and waste pickers, amount of waste plastic completely removed from the ecosystem, and the sustainability of the proposed solution will further be discussed.

569345 Thermochemical Energy Storage Integration with Thermoelectric Power Generation

This paper will investigate the combination of thermochemical energy storage (TCES) with thermoelectric power generation for electricity production and storage in distributed settings. Thermoelectric generation (TEG) of electricity has been heavily studied as a potential method of utilizing solar or waste heat. For low peak power demands in distributed settings, TEGs have the advantage over conventional power generation cycles as they do not require moving parts, large capital investments, or long startup times. Already, thermoelectric generators (TEGs) are being marketed as accessories to camping or wood stoves to charge electronics without any moving parts. However, in many distributed settings, fuels are either unavailable, or in the case of biomass, may require enormous investments in time to gather. Thermal energy storage therefore becomes one possible method of providing the heat input to a TEG device. Although previously, sensible or latent energy storage methods have been coupled with TEG, the possibility of using thermochemical energy storage (TCES) has not been investigated. Although the overall energy efficiency of a TCES + TEG system is lower than a PV + battery storage system, it is expected that thermochemical energy storage offers much cheaper and scalable storage which, depending on the scale, could more than compensate for the cost of the TEG. Presented here is economic motivation for pursuing this unconventional and unlikely route to clean electricity and a preliminary experimental demonstration of a TCES+TEG concept.

571101 Techno-Economic Analysis and Life Cycle Assessment of Decentralized Preprocessing System for Fast Pyrolysis Biorefineries with Blended Feedstocks in the Southeastern USA

Biofuel production has a rapid growth in the past decades. Since 2002, the bioenergy portion contributed by biofuel has reached over five times in the United States and bioenergy was over 5% of the total U.S. energy consumption in 2013.[1] Rapid expansion of biofuel production results in the challenges in the traditional biomass supply chains that have large impacts on the economic effectiveness and production robustness of biorefineries. Traditional biorefinery systems with a single type of feedstock and centralized design may be vulnerable to variations in feedstock cost, quality, and availability. One possible solution is to introduce blended feedstocks and decentralized preprocessing facilities. Many Life Cycle Analysis (LCA) and Techno-Economic Analysis (TEA) have been developed to either investigate the benefits of decentralized biofuel supply chains[2] or blended feedstocks[3]. However, few of them have studied the potential economic and environmental trade-offs of using blended biomass with decentralized preprocessing sites, nor did they quantify the impacts of biomass quality and regional availability, and selection of preprocessing technologies on the overall life-cycle impacts.

This study addressed this knowledge gap by developing TEA and LCA models for biorefineries treating blended feedstocks (i.e., pine residues and switchgrass) with fast pyrolysis. "Depot", the decentralized preprocessing site, was introduced to preprocess biomass for size reduction, mixing, drying, and pelletization. Two different preprocessing technologies were explored: conventional pelleting processing (CPP) and high moisture pelleting process (HMPP).[4] Different scenarios were developed to investigate the impacts of blending ratio of pine and switchgrass, biorefinery and depot capacities, the number of depots, and preprocessing technologies. A Geospatial Information System (GIS) coupled with real-world geographic and biomass data was used to identify the locations of depots and biorefinery along with transportation distances. The TEA model calculated the Minimum Fuel Selling Price (MFSP) based on feedstock costs, the inputs from GIS, and the mass and energy balance in biorefinery from Aspen Plus process simulations. Those data were also used in combinations with the Life Cycle Inventory (LCI) data collected from literature to estimate the cradle-to-gate environmental footprints if different scenarios. The results of TEA and LCA were compared to further understand the trade-offs.

Preliminary results showed that MFSP highly depended on the biorefinery capacity, the blending ratio of feedstocks, and the process pathways of the depot. For biorefinery with a capacity from 1,000 to 2,500 oven dry ton (ODT)/day, MFSP decreased as the capacity increased. The impacts of blending ratio on MFSP were different for biorefineries with and without depots. For biorefinery without depot at 1,000 ODT/day, using a higher ratio of pine residues was more economically beneficial. As the capacity expands, the advantage of using pine residues gradually vanished. It was found that using more switchgrass at 2,000 and 2,500 ODT/day biorefineries resulted in lower MFSP. For biorefineries with depot, MFSP was lower for blending feedstocks than single feedstock type (pine residues). Additionally, introducing the decentralized depots showed the potential to lower the MFSP for biorefinery larger than 2,500 ODT/day. The results also showed that the HMPP preprocessing pathway had lower costs than CPP pathway.

Preliminary LCA results showed that the depot capacities or biorefinery capacities had relatively minor impacts on the life cycle primary energy consumption and Greenhouse Gas (GHG) emissions. Biorefineries, with or without depots, accounted for the major portion of primary energy consumptions. By using real biomass availability data, and investigating the impacts of variabilities in biomass components, preprocessing pathways, depot capacities, and biorefinery capacities, this study provides useful information for a variety of stakeholders for decision making related to investment and design of low-cost and large-scale biorefineries.

References

1. US EIA. Biofuels production drives growth in overall biomass energy use over past decade

https://www.eia.gov/todayinenergy/detail.php?id=15451. Accessed March 15, 2019.

2. Pérez, A. T. E., Camargo, M., Rincón, P. C. N., & Marchant, M. A. Key challenges and requirements for sustainable and industrialized biorefinery supply chain design and management: a bibliographic analysis. *Renewable and Sustainable Energy Reviews*. 2017, *69*, 350-359.
3. Akgul, O., Shah, N., & Papageorgiou, L. G. An optimisation framework for a hybrid first/second generation bioethanol supply chain. *Computers & Chemical Engineering*.2012, *42*, 101-114.
4. Lamers, P., Roni, M. S., Tumuluru, J. S., Jacobson, J. J., Cafferty, K. G., Hansen, J. K., Kenney, K., Teymouri, F. and Bals, B. Techno-economic analysis of decentralized biomass processing depots. *Bioresource technology*. 2015, *194*, 205-213.

576328 Ammonia Decomposition to Produce Hydrogen Using Chemical Looping Approaches: Considerations and Results for High Pressure Operation

A potential solution to reduce transportation costs for H_2 is to use ammonia (energy density: 4.25 kW/L). Ammonia (NH_3) is proposed to be an alternative H_2 carrier, due to its lower transportation costs and it is widely produced on an industrial scale using Haber-Bosch chemistry. From an end-user perspective, if ammonia is used as a H_2 carrier and if H_2 is still used as the primary energy source, there exists a technological need to convert NH_3 back to H_2 in an efficient way. OSU has proposed a thermochemical solids looping based NH_3 to H_2 (ATH) process that can operate at high thermal and H_2 production efficiencies. The ATH system can be operated as a two reactor system or a three reactor system, wherein the first the reducer reactor utilizes an intrinsic O_2 gradient driven by the reduction potential of NH_3 and a metal oxide Fe_2O_3/Fe_3O_4) to efficiently crack NH_3 to a mixture of N_2, H_2 and H_2O. The reduced metal-oxide from the reducer reactor is re-oxidized in a second reactor oxidizer using H_2O as the oxygen source. Depending on the system configuration, the three reactor system adds in a combustor reactor to re-oxidize the metal-oxides. Previous studies have focused on comparing the three reactor system to a two-reactor system and obtaining highest cold-gas efficiencies for each. The earlier chemical looping investigations quantified the thermodynamics of the chemical looping system for atmospheric pressure. So, while the earlier results are promising, it should be noted that various H_2 utilization applications require H_2 at pressures of ~30-50 bar.

This study seeks to quantify the thermodynamic feasibility associated with high-pressure chemical looping applications for Hydrogen production. This study initially quantifies the H_2 production potential of the chemical looping system for higher pressures under isothermal operating conditions. Identified optimal isothermal operating conditions are used for simulating a commercially relevant adiabatic reactor operation. Experimental results are used to provide kinetic parameters to limit design space of the chemical looping system obtained using thermodynamic simulations. The trade-off analysis shows individual and synergistic effects of variables like compression ratios, reactor size, hourly space velocities while using pinch and transshipment type technology for heat exchanger network synthesis. Preliminary results show that operating chemical looping system at higher pressures is feasible and the three reactor system maintains better performance than the two reactor chemical looping system at higher pressures.

568754 Carbonate Fuel Cells: Scaling-up the Magic Carbon Capture Box

To economically sequester or utilize CO_2 from dilute sources such as industrial emissions or power generation, technologies for concentrating CO_2 are required. Unfortunately, commercially available technologies are capital and energy intensive, reducing their effectiveness in limiting emissions. The advantage of the carbonate fuel cell is that it can effectively concentrate CO_2 and simultaneously produce electricity and hydrogen. This example of process intensification takes advantage of the electrochemical reaction to drive a desirable separation. Although carbonate fuel cells have been deployed commercially for power generation, strenuous conditions of carbon capture are much more challenging. A joint development program has been created to reengineer the fuel cell stack for carbon capture, utilizing model-guided and parallel development strategies.

Work on single cells has shown that the use of dilute CO_2 feeds from flue gas sources can have a significant effect on molten carbonate fuel cell behavior beyond increased cathode overpotentials. Specifically, testing has shown that the electrolyte can behave as a dual ion conductor passing both carbonate and non-carbonate ions through the electrolyte. Kinetic testing has helped elucidate mechanistic reaction pathways and suggest that the non-carbonate ions are likely OH^- formed from the catalysis of H_2O and O_2. In addition, mass transport effects are found to be of increased importance at these conditions and a leading factor in controlling fuel cell ionic selectivity. These learnings highlight the need for several design modifications which can lead to improved fuel cell performance in carbon capture applications.

A kW scale technology stack with commercially sized repeat components is being used to evaluate the scale-up challenges of mass transfer and thermal management. Insights from the lab scale fundamentals research have been applied to design a cathode current collector that reduced the CO_2 mass transfer limitations in the fuel cell cathode. With a redesigned current collector, the stack ionic selectivity was improved by >15% while simultaneously increasing voltage by >40 mV. Additionally, because of the exothermic nature of the electrochemical reactions and electrical resistivity, thermal management is critical. Endothermic steam methane reforming is used to heat balance the fuel cell stack and provide hydrogen fuel. Carefully co-locating the heat generation and consumption reactions is critical to reducing thermal gradients rather than exacerbating them. A modeling approach was utilized to design a methane reforming catalyst pattern that could reduce thermal gradients by ~40% within the fuel cell stack and allow for improved electrochemical performance.

The carbonate fuel cell carbon capture technology is able to reduce parasitic energy demands of carbon capture while actually co-producing hydrogen and electricity. The deployment of this unique reactive separation technology, however, relies on careful engineering to properly manage the heat and mass transfer during scale-up. Fundamentals based models have been key in guiding the scale up by highlighting the key mass and heat transfer issues. Going forward, the model guided design approach will be key in matching process conditions and hardware design to specific carbon capture applications.

570292 Building Block-Based Work and Heat Exchanger Network Synthesis (WHENS)

Energy conservation has been a long-standing grand challenge in chemical industry. Efforts in heat exchanger network synthesis facilitate saving in energy consumption and total annual cost [1-2]. Recently, work and heat exchanger network synthesis (WHENS) has emerged as a promising research area to capture the interplay of pressure and temperature manipulation operations [3-5]. Current optimization-based approach for WHENS leverages on superstructure representations such as state-space representation [6], multi-stage superstructure [7] and representations involving heuristics [8]. However, there exist several limitations for these superstructure representations addressing WHENS. Firstly, equipment configurations need to be postulated based on existing knowledge of unit operations, engineering experience and heuristics. Secondly, classic superstructure representations fail to discover novel pathways with process intensification. Process intensification refers to significant reduction of equipment sizes, waste generation, and increase of productivity [9,10]. In the context of WHENS, multi-stream heat exchanger is a relevant example that reduce the number of equipment while improving the overall performance of WHENS.

To this end, we provide a new representation approach to capture many process configurations of WHENS with innovative pathways [11]. The new representation approach centers around two fundamental elements of abstract building blocks. Firstly, the block interior is used to represent splitting, mixing, utility cooling and utility heating of individual streams. Secondly, the block boundaries between adjacent blocks allow energy flow in the form of heat and work. An unrestricted boundary does not allow pressure and composition changes between adjacent blocks. A semi-restricted boundary is assigned with expansion/compression operations connected through either common (integrated) or dedicated (utility) shafts. A completely restricted boundary allows heat flow driven by temperature gradient across the boundary. We demonstrate the rich process information embedded in the proposed representation through literature examples. The novelty of this approach is that there is no need to specify the stage number as requested by the classic approach. Besides, the proposed representation approach allows automatic generation of numerous work and heat exchanger network with intensification opportunities. We formulate this building-block based WHENS as a mixed-integer nonlinear optimization (MINLP) problem with the objective as minimizing total annual cost. We use several literature examples on liquified energy chain to demonstrate the capability of the proposed synthesis approach.

References:

[1] Furman, K.C.; Sahinidis, N.V. A critical review and annotated bibliography for heat exchanger network synthesis in the 20th century. *Industrial & Engineering Chemistry Research*. 2002, 41, 2335–2370.

[2] Nair, S. K., & Karimi, I. A. Unified Heat Exchanger Network Synthesis via a Stageless Superstructure. *Industrial & Engineering Chemistry Research*. 2019. DOI: 10.1021/acs.iecr.8b04490.

[3] Yu, H.; Fu, C.; Vikse, M.; Gundersen, T. Work and heat integration—A new field in process synthesis and PSE. *AIChE Journal*. 2018.

[4] Fu, C.; Vikse, M.; Gundersen, T. Work and heat integration: An emerging research area. *Energy*. 2018, 158, 796–806.

[5] Huang, K.; Karimi, I. Work-heat exchanger network synthesis (WHENS). *Energy*. 2016, 113, 1006–1017.

[6] Wechsung, A.; Aspelund, A.; Gundersen, T.; Barton, P.I. Synthesis of heat exchanger networks at subambient conditions with compression and expansion of process streams. *AIChE J.* 2011, 57, 2090–2108.

[7] Nair, S.K.; Nagesh Rao, H.; Karimi, I.A. Framework for work-heat exchange network synthesis (WHENS). *AIChE J.* 2018, 61, 871–876.

[8] Uv, P.M. Optimal Design of Heat Exchanger Networks with Pressure Changes. Master's Thesis, NTNU, Trondheim, Norway, 2016.

[9] Tian, Y.; Demirel, S.E.; Hasan, M.M.F.; Pistikopoulos, E.N. An Overview of Process Systems Engineering Approaches for Process Intensification: State of the Art. *Chem. Eng. Process.-Process Intensif.* 2018, 133, 160–210.

[10] S.E. Demirel; J. Li; M.M.F. Hasan. Systematic Process Intensification. *Current Opinion in Chemical Engineering*, 2019, https://doi.org/10.1016/j.coche.2018.12.001.

[11] J. Li; S.E. Demirel; M.M.F. Hasan. Building Block-Based Synthesis and Intensification of Work-Heat Exchanger Networks (WHENS). 2018. *Processes*, 7(1), 23.

567580 Photocatalytic Synthesis of Benzylic Amines Using Metal-Free Dyes: Investigating the Reaction Performance with a Spinning Disc Reactor and a Microchannel Reactor

-

Solar photocatalysis has grown as a green alternative for the synthesis of fine organic compounds. Recently, metal-free dyes have gained interest as visible light photocatalysts as they are more environmentally benign than transition metals, they are cheap and they have long excited-state lifetimes.[1] Eosin Y (characteristic peak 539 nm) is an anionic dye, which has efficiently catalysed a range of reactions including cyclisation, halogenation and redox reactions.[2,3] Despite the advances in the lab-scale synthesis, there is a lack of knowledge on how to scale up these visible light photocatalytic reactions for industrial implementation. To absorb light efficiently, thin films and large surface areas are needed. To date, most photocatalytic reactors have been designed based on enclosed UV-light and are therefore not suitable for visible light. Neither are traditional reactors, such as batch or packed reactors, and thus new process intensification (PI) reactors need to be designed, optimised and evaluated to maximise the light efficiency and to overcome the mass transfer limitations.

The aim of this study is to evaluate two different types of thin film/large surface area reactors (Fig 1) for the intensification of visible light photocatalysis: (i) Spinning Disc Reactor (SDR) and (ii) Microchannel reactor (MCR). A specially designed solar light simulator (40cm*40cm) was used to carry out the oxidative homocoupling of benzylamine, chosen as a model reaction.

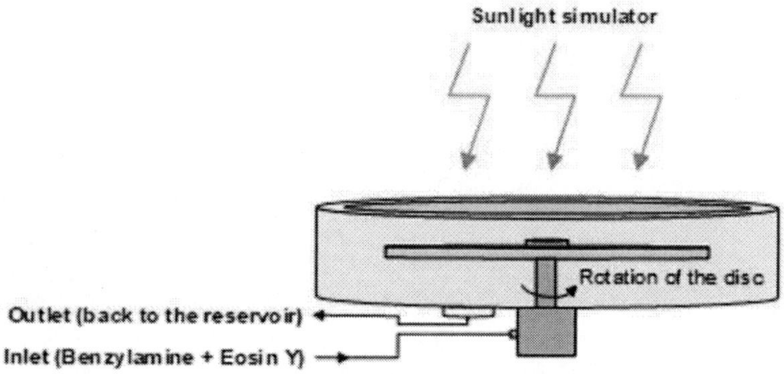

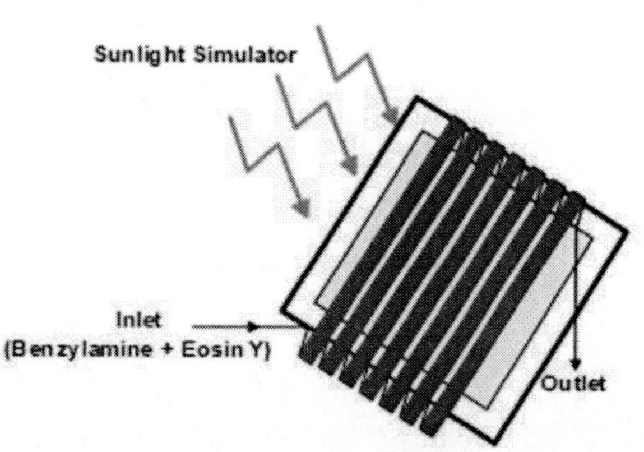

a) b)

Figure 1: Schematic representation of a) Spinning Disc Reactor (SDR); b) Microchannel reactor (MCR).

A range of conditions with low and high substrate concentrations and catalyst loadings were tested with both configurations. As opposed to batch reactors, the photon-transfer limitation was overcome due to the creation of thin films by the centrifugal acceleration in the SDR, and the large surface area per volume ratio in the MCR. Moreover, thin films and high surface areas reduce the effect of the opacity of the solutions, as the light is able to penetrate into the mixture. Therefore, high catalyst loadings and amine concentrations can be used to improve the productivity of the reaction. In Fig 2a, it can be observed that for the MCR, at a high amine concentration, higher catalyst loadings increased the conversion. The productivity also increased as, for example, with 6 mol% Eosin Y and 5 mM initial concentration of amine, 0.41 mmol of amine reacted to form the product whereas with 15 mM, 1.21 mmol of amine reacted.

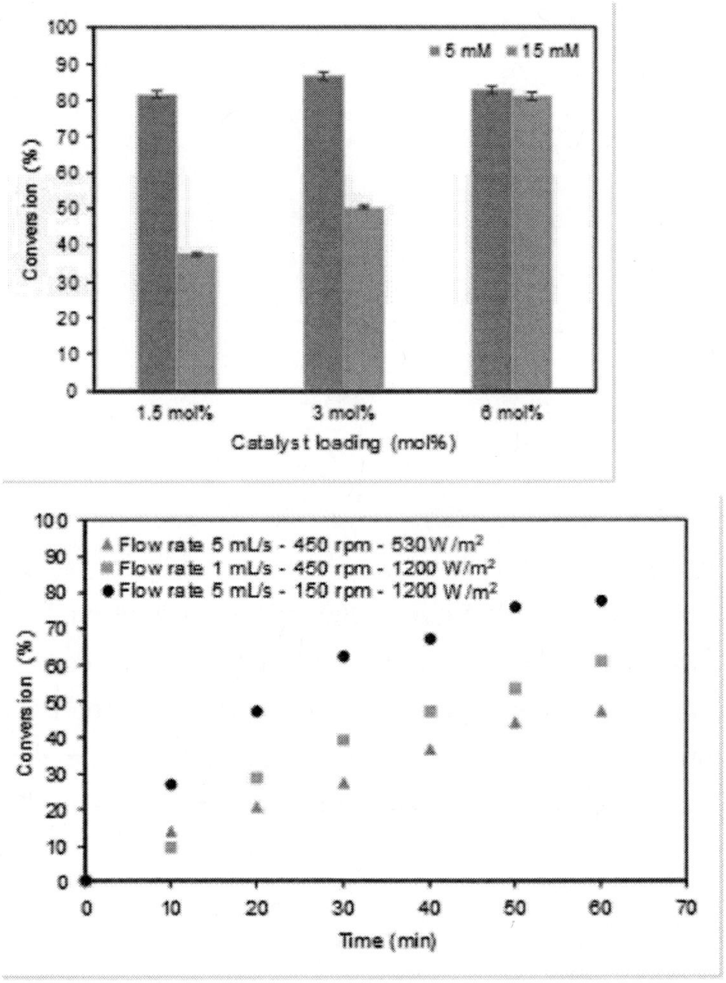

a) b)

Figure 2: a) Effect of catalyst loading and amine concentration using the MCR; b) Effect of flow rate and spinning speed using the SDR.

One of the most significant factors in the performance of the SDR, similarly to the MCR, is the interaction between the flow rate and the light intensity (Fig 2b). This will determine the time that the catalyst spends under the light, in the catalyst-excited state. High flow rates permit a faster recirculation of the solution and therefore, the reaction mixture is on top of the disc more times during a specific reaction time. Due to the thin films created by the SDR (film thickness of 0.09 – 0.45 mm), the light is able to penetrate into the solution and the spinning speed does not greatly influence the conversion.

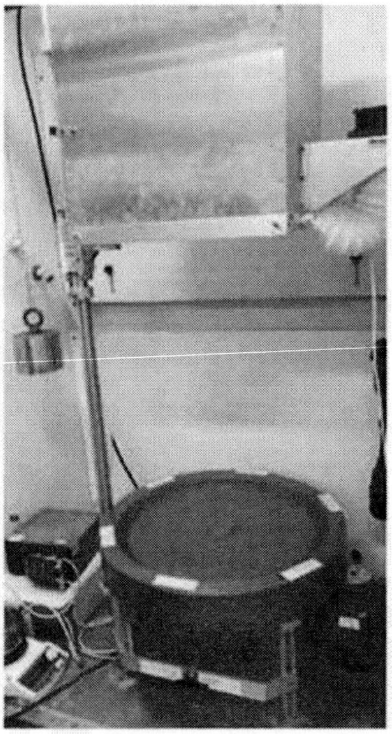

Figure 3. Setup of the solar light simulator and the SDR.

These studies have demonstrated that the MCR and the SDR have the potential to enable intensification of reactions with commercial importance. Further investigations are being carried out with these reactor configurations and with a batch reactor. Moreover, the performance of the SDR (Fig 3) is being investigated with an immobilised catalyst, which creates a process where the catalyst can be simply recovered and reused. In summary, the combination of metal-free dyes and novel visible light photocatalytic reactors have shown to be a solution to maximise the light efficiency, overcome the mass transfer limitations and create a more sustainable process to synthesise fine organic chemicals.

REFERENCES

1 J. R. Xin, J. T. Guo, D. Vigliaturo, Y. H. He and Z. Guan, *Tetrahedron,* 2017, **73**, 4627-4633.

2 D. Prasad Hari and B. König, *Chem. Commun,* 2014, **50**, 6688-6699.

3 J. D. Tibbetts, D. R. Carbery and E. A. C. Emanuelsson, *ACS Sustain. Chem. Eng.*, 2017, **5**, 9826-9835.

564044 Oxidation of L-Methionine in a Rotor-Stator Spinning Disc Reactor

Oxidation of L-Methionine in a Rotor-Stator Spinning Disc Reactor

Arnab Chaudhuri[1] , Koen Kuijpers[1], Parimala Shivaprasad[1,2], Emma Anna Carolina Emanuelsson[2], Timothy Noël[1], John van der Schaaf[1]

[1]Department of Chemical Engineering, TU Eindhoven, 5612 AZ, Eindhoven, Netherlands

[2]Department of Chemical Engineering, University of Bath, BA2 7AY, Bath, United Kingdom

Process intensification (PI) is a promising pathway to achieve sustainable chemical production while also improving the safety of traditional chemical processes [1]. The rotor-stator spinning disc reactor (rs-SDR) is one such PI reactor which consists of a rotating disc enclosed by a stator. Reaction intensification is caused by the high shear force on the feed as a result of the velocity gradient between the rotor and the stator; resulting in high rates of mass and heat transfer [2]. This leads to rs-SDR being a particularly interesting reactor for photochemical applications where photon transfer and mass transfer of the activated species are often limited (e.g. for high concentration). In this study, we demonstrate the potential of a photo-rs-SDR for applications towards photochemical reactions. The oxidation of L-methionine with methylene blue as the catalyst (Figure.1) was chosen as the model reaction. The gas-liquid mass transfer limitations[3] of this systems makes it particularly applicable for the photo-rs-SDRs and demonstrates the potential for improvements in productivity.

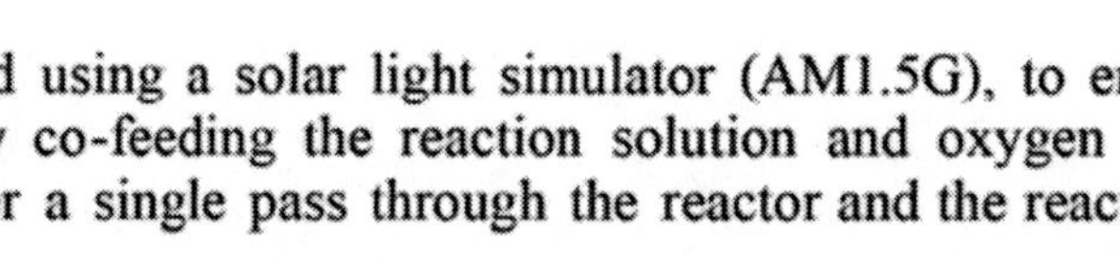

Figure 1: Reaction scheme

The rs-SDR was illuminated using a solar light simulator (AM1.5G), to ensure no light limitations. The reaction was carried out by co-feeding the reaction solution and oxygen at the bottom of the reactor. Samples were collected after a single pass through the reactor and the reaction conversion was measured using HPLC.

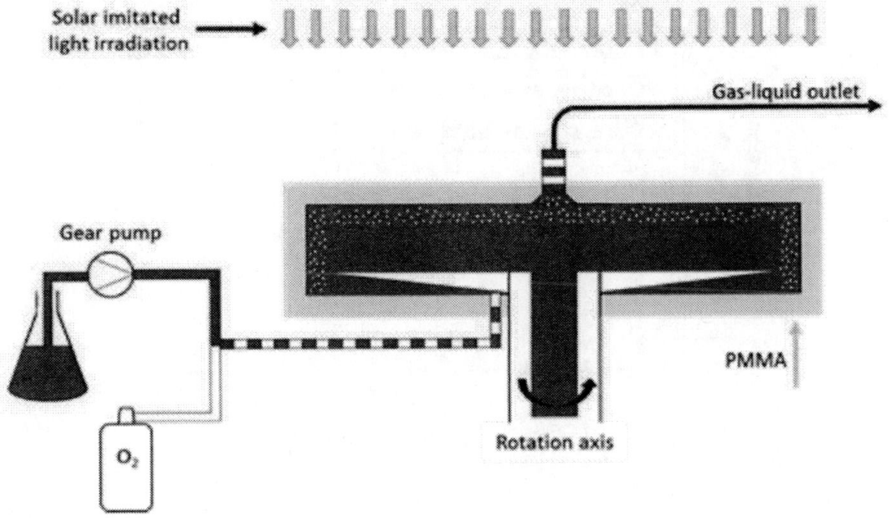

Figure 2: Schematic representation of the reactor set-up

A few of the results obtained are presented in this abstract. Two different configurations were used for the photo-rs-SDR, one with illumination on the dispersed region and one with illumination on the thin film side. As illustrated in Fig 3, the dispersed region seemed to provide better conversions. This is most likely since the mass transfer rates are higher in this region and this allows for more oxygen in the liquid phase, leading to improvements in reaction rates. As the liquid flow rates are increased for constant gas flowrates, the conversions obtained are quite low regardless of the reactor configuration.

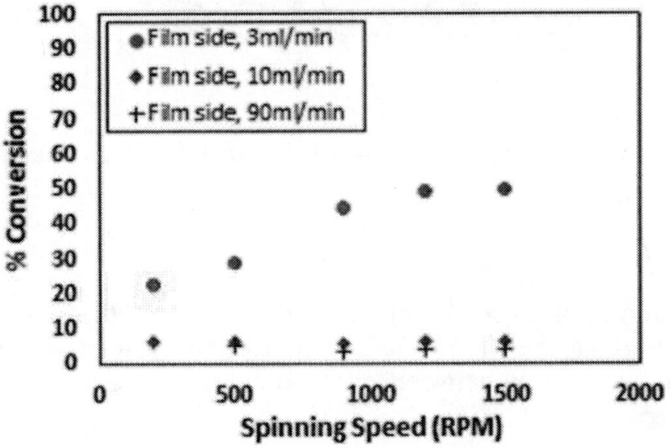

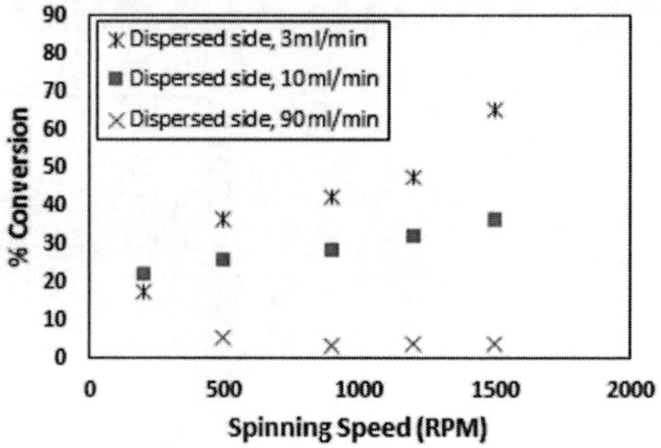

Figure 3: Effect of spinning speed and gas-liquid flowrate on reaction conversion

Figure 4: Visual studies of flow regimes in the photo rs-SDR indicated that uniform distribution of gas could be achieved with increased rotation speed.

Visual studies (Fig 4) of the reactor illustrated that for 1:1 gas to liquid flow ratios, uniform distribution of gas could be achieved throughout the illuminated section of the reactor for the dispersed configuration. This further indicates that the contact between gas and liquid phases in this region was quite high and improved with rotation speeds.

Quite high conversions were achieved in this reactor with residence times of 10 mins. This indicates that the productivity of L-methionine oxidation can be improved. The effect of rotation speed, gas-liquid flowrate and catalyst concentration were investigated on the reaction conversion. We have also investigated these parameters under more light limiting conditions. Additionally, the results were compared to two conventional reactors which are currently used in photochemistry: batch and microflow.

To the best of the authors' knowledge, this is the first study of a successful synergistic reaction system using photochemistry with the rs-SDR. We are currently investigating the potential of the photo rs-SDR for targeted applications in pharmaceuticals and the fine chemical industries.

References

[1] Vlachos, D.G. and S. Caratzoulas, Chemical Engineering Science **2010**, 18-29.

[2] Meeuwse, M., J. van der Schaaf, and J.C. Schouten, *AIChE Journal* **2012**, p. 247-255.

[3]Emmanuel, N., Mendoza, C., Winter, M., Horn, C. R., Vizza, A., Dreesen, L,& Monbaliu, J. C. M., *Organic Process Research & Development*, **2017**,p. 1435-1438.

569946 Development of Sorption Enhanced Reactive Processes for the Sustainable Production of Solketal through the Ketalization of Glycerol

Development of sorption enhanced reactive processes for the sustainable production of Solketal through the ketalization of glycerol

Several research groups around the world have recently identified Solketal as one of the most promising solution towards the valorization of glycerol which has become one of the major topics to ensure the sustainability of the entire biodiesel production chain, since this polyol is obtained as a by-product of the Biodiesel production representing approximately 10 wt% of the total mass of diesel produced. Solketal presents interesting properties as diesel additive which would allow its addition to the final biodiesel formulation not only increasing the quality and the amount of diesel produced but also reducing the overall process by-products formation.

For the synthesis of Solketal, glycerol is commonly reacted with acetone in the presence of an acid catalyst, forming water as by-product [1]. Following the general "Green Chemistry" guidelines, the use of heterogeneous catalysts as ion-exchange resins, zeolites and some silica base materials has been proposed [2]; however, most of the research published focused in batch reactor studies for the optimization of the reaction conditions. In another approach, multifunctional reactors have also been proposed for the production of solketal, which can represent a sustainable alternative for the production of this compound, due to their ability to overcome the thermodynamic limitations associated with this chemical reaction by continuously removing one of the products from the reaction media, which will also result in an increase of the process productivity and in a reduction of its capital cost. In fact, Clarkson et al. [3] and Roldan et al. [4] have proposed the use of reactive distillation and membrane reactors for this purpose, respectively. Although the results presented where satisfactory in terms of glycerol conversion and solketal yield (typically above 90%), it is commonly acknowledged that distillation processes present a high energy demand and zeolite membranes lack long term stability. In this context, the present work represents an innovative approach for the production of solketal making use of sorption enhanced reactive processes, which can outcome the previously studied multifunctional reactors performances based on the high purities and productivities attained by adsorptive-reactors in similar systems [5] and by the process energy efficiency and lower operating temperatures.

A previous study concluded that Amberlyst-35 and ethanol were the most suitable stationary and mobile phases for the synthesis of solketal through a sorption enhanced reactive process, respectively, and fundamental adsorption and reaction data were determined through independent set of experiments providing essential knowledge for the definition of some of the most relevant process operating conditions.

The synthesis of solketal was then experimentally performed in a fixed bed adsorptive reactor. The maximum conversion achieved in a transient state by feeding an equimolar mixture of glycerol and acetone diluted in 50% of ethanol (at 313 K and 5 mL/min) was 60%, while the steady state conversion achieved under these conditions was only 44%. Nevertheless, this value is still close to the reaction equilibrium conversion which is approximately 50%. By changing the reactants molar ratio in the feed mixture from 1:1 to 2:1 (with an excess of acetone) the maximum conversion attained reached 80% in a transient state and the equilibrium conversion was achieved at steady state. These results demonstrate the potential of sorption enhanced reactive processes for the synthesis of solketal. Moreover, it was possible to develop a mathematical model that was able to accurately describe the experimental results obtained in the fixed bed adsorptive reactor experiments considering isotherm operation and an axially dispersed plug flow at a constant bed length and porosity. The non-idealities of the mixture were accounted for by determining the activity coefficients through the UNIFAC method and the internal and external mass transfer coefficients between the liquid and the solid phase were estimated and lumped in a global mass transfer coefficient.

Nevertheless, the fixed bed adsorptive reactor productivity is known to be limited requiring, at the same time, the use of significant amounts of solvent for regeneration purposes, and it would not allow the implementation of a continuous production process. Therefore, the production of solketal in a Simulated Moving Bed Reactor (SMBR) was studied, as this chromatographic reactor overcomes all the aforementioned drawbacks. For that purpose, glycerol and acetone were used as reactants and were introduced in the unit through the feed stream while ethanol was used as desorbent. As solketal is the less retained species it will be collected in the raffinate stream while water, the most retained one, will be collected in the extract stream. Consequently, water is selectively removed from the reaction media leading to the displacement of the thermodynamic reaction equilibrium and considerably increasing the reaction conversion. To evaluate the performance of this technology, a comprehensive simulation study was performed, based on the True Moving Bed Reactor model approach reported elsewhere [6], to optimize the design variables associated with the SMBR operation following to main approaches: using a conventional 4 sections SMBR and feeding a 1:1 mixture of glycerol and acetone, and using a non-conventional 5-sections SMBR in which pure glycerol and pure acetone are fed through independent streams at different columns of the unit (typically feeding acetone closer to the extract port and glycerol closer to the raffinate port to promote the counter-current movement of both of these species based on their adsorption selectivity).

The results for the conventional 4-sections SMBR demonstrated that the purity (in an eluent-free basis) attained in the raffinate stream was 90% which is relatively low compared with the values obtained for similar acetalization reactions performed in SMBR (which are typically comprised between 95% and 99% [5]). The low equilibrium conversion together with the low adsorption selectivity between acetone and solketal when using Amberlyst-35 as catalyst/adsorbent and the significant mass transfer resistances associated with this system represent the most likely causes behind the technology limited performance. On the other hand, the 5-section SMBR unit was able to achieve a solketal purity of 95% without compromising the productivity of the process and without leading to any increase in the desorbent consumption. The main factor contributing for this result was the counter-current contact between the two reactants promoted by feeding them independently at different locations within the unit. As the more retained reactant (glycerol) was fed closer to the raffinate port and the less retained one (acetone) was fed closer to the extract port, a considerable increase in the overall reaction rate and reactants conversion was observed throughout the unit. Additionally, by feeding acetone to a column placed at a larger distance from the raffinate collection port, it was possible to enhance the separation between solketal and acetone.

Therefore, it is possible to conclude that sorption enhanced reactive processes represent a potential alternative for the sustainable production of solketal using glycerol as raw material.

References

[1] Bruchmann, B., Haberle, K., Gruner, H. and Hirn, M., Preparation of cyclic acetals or ketals, US5917059 A, 1999.

[2] Trifoi, A.R., Agachi, P.S. and Pap, T., Glycerol acetals and ketals as possible diesel additives. A review of their synthesis protocols. Renewable and Sustainable Energy Reviews, 2016. 62: p. 804-814.

[3] Clarkson, J.S., Walker, A.J. and Wood, M.A., Continuous Reactor Technology for Ketal Formation: An Improved Synthesis of Solketal. Organic Process Research & Development, 2001. 5(6): p. 630-635.

[4] Roldán, L., Mallada, R., Fraile, J.M., Mayoral, J.A. and Menéndez, M., Glycerol upgrading by ketalization in a zeolite membrane reactor. Asia-Pacific Journal of Chemical Engineering, 2009. 4(3): p. 279-284.

[5] Rodrigues, A.E., Pereira, C.S.M. and Santos, J.C., Chromatographic Reactors. Chemical Engineering & Technology, 2012. 35(7): p. 1171-1183.

[6] Rodrigues, A.E., Pereira, C.S.M., Minceva, M., Pais, L.S., Ribeiro, A.M., Ribeiro, A., Silva, M., Graça, N. and Santos, J.C., Simulated Moving Bed Technology: Principles, Design and Process Applications. 2015: Elsevier Science.

578342 Gradual Synthesis of Heat Exchanger Networks Under Economic, Environmental and Safety Considerations

The heating and cooling needs in the chemical process industry are typically covered using external utilities. Process intensification in the form of energy integration among process streams is an effective tool to achieve more energy-efficient designs. We address in this work the design of heat exchanger networks (HENs), but instead of developing a design aiming for a full energy integration we examine the effect of performing gradual process integration by adding process-process exchangers until the complete integration is achieved to analyze the step-by-step effect of such a strategy.

The problem starts by considering m hot and n cold process streams to be integrated, for which targets for minimum energy consumption are determined through pinch point techniques. Then, the individual contribution of each heat exchanger used to integrate energy is analyzed, and the compromise between the investment cost and the reduction of utility consumption is evaluated. To complement the analysis environmental and safety factors are considered for each of the networks that result from a structure with no energy integration to a final one that achieves complete integration that meets the minimum energy consumption targets.

The strategy for gradual energy intensification is based on the heat duties required by each exchanger. Therefore, we start with the implementation of the heat exchanger that shows the largest duty, for which its size, cost, and contribution towards the objective of minimum energy consumption are determined. The effect on the environmental and safety indicators is also assessed. This procedure is applied sequentially until full energy integration is accomplished.

The results are analyzed through several metrics. From economic terms, each integrated structure is evaluated taking into account the saving generated through energy integration and the investment cost for the heat exchanger(s). If marginal return on investment is of interest, it is shown that partial integration provides a better policy. Environmental and safety indicators are also considered. The environmental impact of each partial network is evaluated in terms of the reduction of CO_2 equivalent emissions that are obtained as heat exchangers are added to the network. Safety aspects are evaluated through the Inherent Safety Index for Shell and Tube Heat Exchangers (ISISTHE)[1], which rates inherent safety considering pressure, heating value, combustibility and an adjusted temperature difference.

The procedure is applied to several case studies ranging from systems with few streams to the aromatics plant problem.[2] By considering the three factors (economic, environmental and inherent safety) it is shown how one can identify the level of integration at which an optimum point that meets the best compromise among them is obtained.

[1]Pasha, M.; Zaini, D.; Shariff, A. M. Inherent safer design for heat exchanger network. J. of Loss Prevention in the Process Industries. 2017, 48, 55-70.

[2]Linnhoff, B.; Ahmad, S. Cost optimum heat exchanger networks-1. Minimum energy and capital using simple models for capital cost. Computers Chem. Engng. 1990, 14, 7, 729-750.

562229 Sensor Fusion with Irregular Sampling and Varying Measurement Delays

In multi-sensor fusion, several sources of information are combined in order to increase the estimation quality for the quantity of interest. This activity finds many applications from tactical missile defense to self-driving cars and the estimation of variables difficult to measure such as concentrations in chemical processes (Gao & Harris 2002, Kumar et al. 2007, Safari et al. 2014). In these situations, it is common to resource to laboratory analysis that provide more accurate measurements, but usually at slower sampling rates and with significant delays. Therefore, soft sensors and online analyzers are often introduced in the process to provide more frequent and updated measurements (Fatehi & Huang 2017).

Since Bar-Shalom & Campo (1986) introduced track-to-track fusion (TTF), several methods have been proposed based on the use of Kalman filter (KF) for sensor fusion. TTF makes use of a KF to track each sensor and fuses the sensors' state estimates into a new state estimate using a static linear estimation equation (Chen et al. 2003). Gao & Harris (2002) reviewed measurement fusion processes and proposed a modified track-to-track fusion (MTF). However, these methodologies rely on first-principle models and do not capitalize on prior knowledge about the relative quality of the sensors. Therefore, in this work we present a KF based sensor fusion scheme that integrates data collected from a dynamic soft sensor, an online analyzer and a laboratory analyzer. Fusion occurs in the absence of a first-principle model and accommodates a less reliable (but more frequent) online analyzer recurring to laboratory measurements. This scheme has the capability for managing asynchronous sampling rates and detecting faulty measurements while reducing false alarms and missed identifications.

Two fusion schemes are proposed and outlined: a more classic tracked Bayesian fusion scheme (TBF) and a novel modification of the track-to-track algorithm, designated bias-corrected track-to-track fusion (BCTTF). When comparing these two schemes, the new BCTTF has shown to be better in terms of prediction performance and alarm identification sensitivity. This algorithm produces also a less noisy signal. The analysis of the figures of merit leads us to recommend the use of BCTTF as a fusion algorithm under multi-rate sensor fusion conditions.

The presented case study arises from a real world chemical plant and comprises two trayed distillation columns for product purification in a Dow production facility. Process measurements include pressures, temperatures and flow rates across the columns, as well as, reflux ratio and online analyzer readings from upstream reactors. Quality variables regard the concentrations of impurity components, say A and B. These are sampled every hour by an online analyzer and a soft sensor. Furthermore, every 12 hours a laboratory sample is grabbed for analysis (Lu & Chiang 2018). By applying the BCTTF algorithm and analyzing the results, we could confirm its good performance and infer that the laboratory sampling rate can be reduced to 1 sample every 48 hours without significant loss of prediction power. This allows for a meaningful cost reduction in laboratory material and hours of work. Finally, an Emergency Laboratory Sampling Plan is outlined to cope with possible faults that may put the online analyzer out of service and to have better control over the process during an abnormal operation time period.

References

Bar-Shalom Y, Campo L. The effect of the common process noise on the two-sensor fused-track covariance. *IEEE Transactions on Aerospace and Electronic Systems*. 1986;22:803-805.

Chen H, Kirubarajan T, Bar-Shalom Y. Performance limits of track-to-track fusion vs. centralized estimation: Theory and application. *IEEE Transactions on Aerospace and Electronic Systems*. 2003;39:386-398.

Fatehi A, Huang B. Kalman filtering approach to multi-rate information fusion in the presence of irregular sampling rate and variable measurement delay. *Journal of Process Control*. 2017;53:15-25.

Gao J, Harris C. Some remarks on Kalman filters for the multisensor fusion. *Information Fusion*.

2002;3:191-201.

Kumar M, Garg DP, Zachery RA. A method for judicious fusion of inconsistent multiple sensor data. *IEEE Sensors Journal*. 2007;7:723-733.

Lu B, Chiang L. Semi-supervised online soft sensor maintenance experiences in the chemical industry. *Journal of Process Control*. 2018;67:23-34.

Safari S, Shabani F, Simon D. Multirate multisensor data fusion for linear systems using Kalman filters and a neural network. *Aerospace Science and Technology*. 2014;39:465-471.

562176 Knowledge-Based Interactive Preprocessing of Process Data

Zhong-Yan Liu[1], Min Hong[1], Min-Yi Peng[1], Yi Heng[2,3,4]*, (1) School of Chemical Engineering and Technology, Sun Yat-sen University, Zhuhai, China, (2) School of Data and Computer Science, Sun Yat-sen University, Guangzhou, China, (3) Guangdong Province Key Laboratory of Computational Science, Guangzhou, China, (4) National Supercomputer Center in Guangzhou (NSCC-GZ), Guangzhou, China

*** Corresponding author. E-mail address: hengyi@mail.sysu.edu.cn (Yi Heng)**

Keyword: Data Preprocessing, Workflow Support, Knowledge Base

How to preprocess data with advanced analysis algorithms and leverage it into applications is a crucial task in practice from a holistic viewpoint [1]. To this end, distributed data processing has been proposed to deal with a wide range of industrial applications [2]. On the one hand, Jashapara et al. considered well-established information technologies such as databases and data mining for discovering knowledge [3-5]. On the other hand, the workflow used for processing real-life data can be created by tools, methods, and standards that provide better support in the domain of knowledge management, modeling, simulation, validation and other aspects of designing, testing, and commissioning [6-8]. Meanwhile, the amount of process data obtained from engineering has been significantly increasing. Personalized and efficient data preprocessing encountered challenges to develop next-generation intelligent systems, such as maintaining the quality, accessibility and traceability of data, reviewing and reusing knowledge as well as providing useful workflows for practical applications.

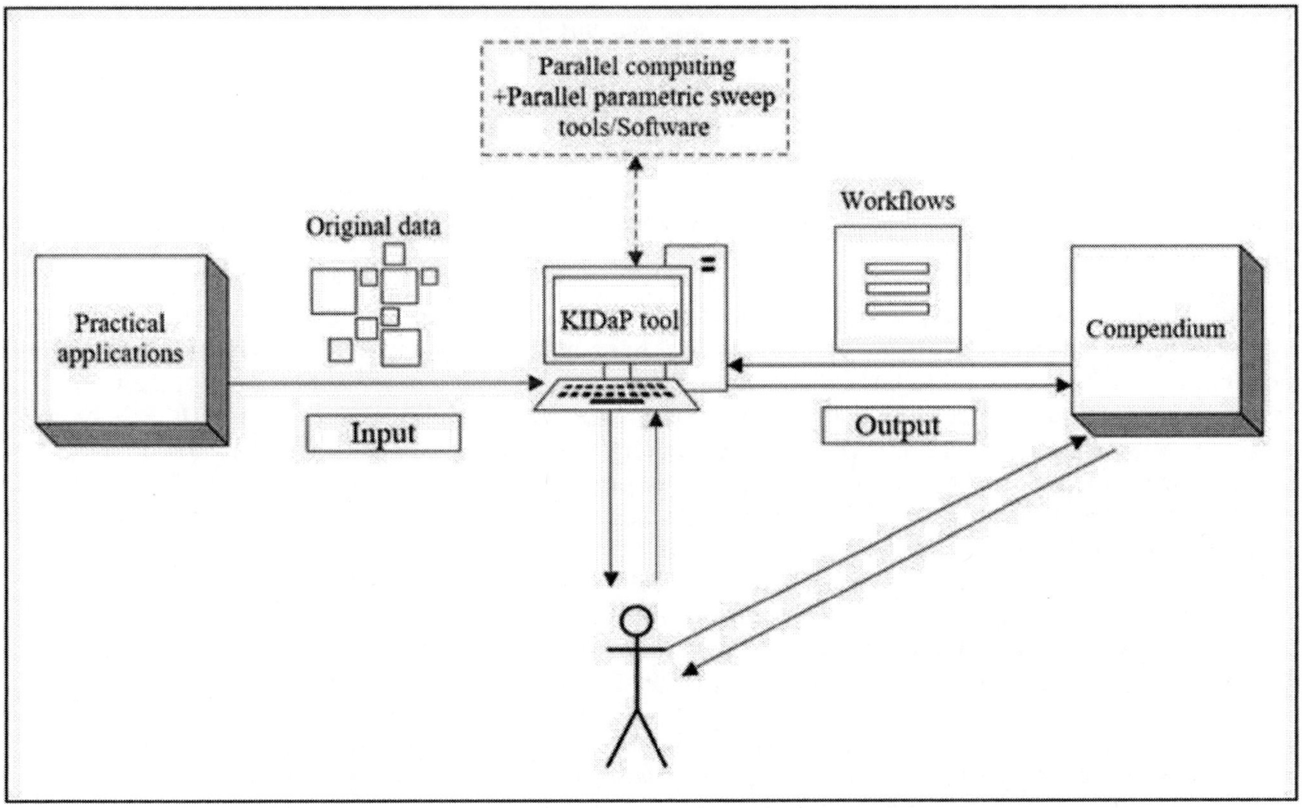

Figure 1. Overview of the KIDaP management framework

In this work, taking an application instance of boiling experiments as an example, we consider a type of management framework that enables knowledge-based big data preprocessing (KIDaP) for practical tasks in a systematic and efficient way (see Fig. 1). The KIDaP management framework was proposed by Heng, Theissen and Soemers [9]. It consists of two main components: a compendium of workflows that summarizes algorithms and methods for massive data processing and a knowledge-based interactive open source software tool KIDaP that has direct access to the compendium. Software user interface is shown in Fig. 2.

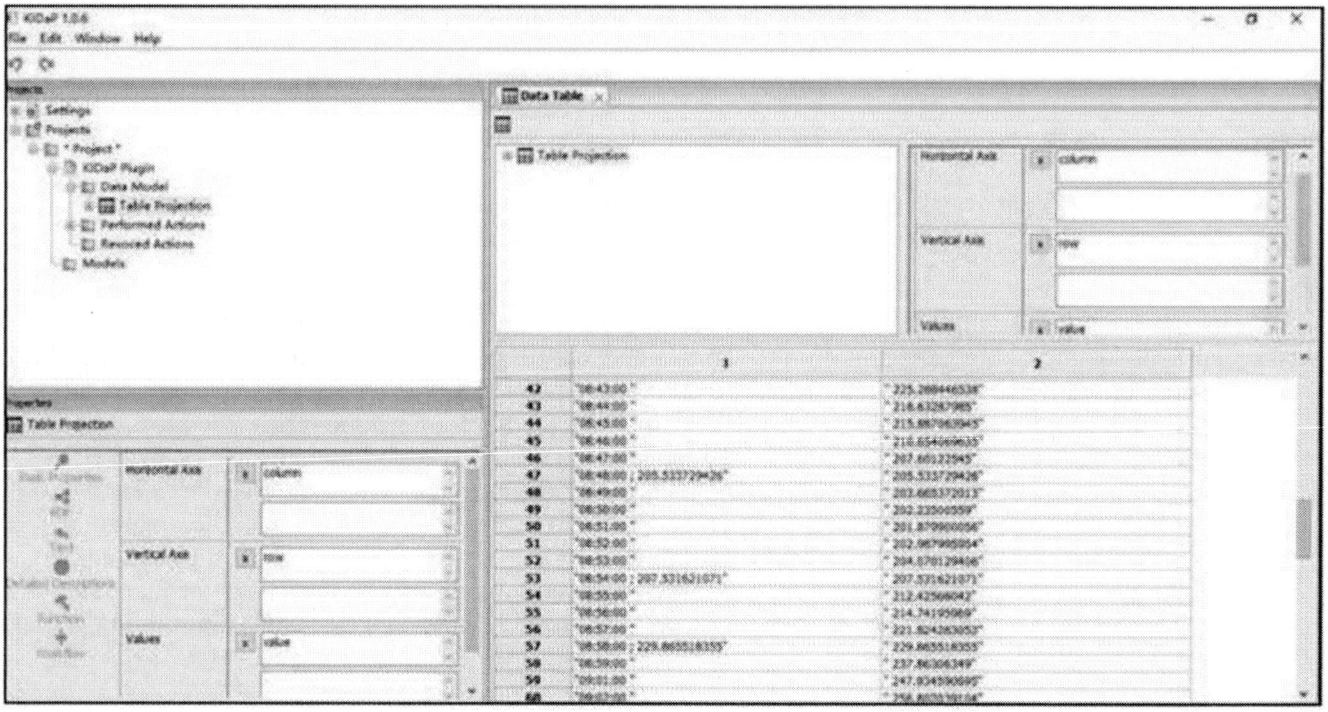

Figure. 2. Software user interface of KIDaP

This knowledge-based management framework is supposed to be used to guide the user to automate repetitive tasks with workflow support, which is open for future extensions in other fields of science and engineering. By using the software KIDaP, users can specify the input format of their data, design and execute a workflow. The use of different processing tools within the platform is available while the designed workflow and related single steps can be automatically recorded so that the data and the actions can be traced easily at a later time.

The data preprocessing task considered in this work arises from a pool boiling experiment conducted by our collaboration partner. It is known that boiling heat transfer has many applications and can be enhanced by micro/nanostructured surfaces. In our collaboration partners' work, wang et al. fabricated micro/nano bi-porous surfaces, taking into account both optimal cavity size and wettability modification [10]. Firstly, a micro/nano bi-porous copper surface with abundant micro pores whose diameters are fabricated in the optimal range for cavity size according to previous study from other researchers [11] is considered. Secondly, a prior development by wettability modification was made to further enhance the boiling heat transfer on the micro/nano bi-porous copper surface. Boiling heat transfer performance of the samples fabricated in these two steps were tested with water at atmosphere pressure respectively [10]. Through the performance test, original experimental temperature time series data are obtained. They manually processed these massive data step by step in order to obtain the boiling performance curve. According to the data analysis results, the mechanism of the boiling heat transfer can be further analyzed.

Here, we are going to use the proposed knowledge-based management framework to preprocess the original data. As shown in Fig. 3, the data processing work-process is divided into five tasks, of which task 4 can be optionally selected. Each task applies a different algorithm or a single step, so that it can process data efficiently. At first, we import the original data (*.CSV) into the KIDaP tool and perform task 1 (prepare data structure) by using the KIDaP built-in functionalities. Then, task 2 (data screening) is executed with an equal-width random selection. In task 3 (data transformation), a series of numerical algorithms (such as Fourier 1-D conduction equation) are used to analyze the data. Task 4 (data integration) is optional, and will be performed when necessary. And then, task 5 (data visualization) will be executed as a final step. The drawing module is called to plot the data, for further analysis and study of the boiling heat transfer mechanism. Compared with manual processing of data, the proposed knowledge-based management framework is supposed to solve the problems of excessive data volume, numerous data processing steps, efficiently process and record the workflows as well as implement parallel parametric sweep, etc. As a long-term goal, the proposed knowledge-based management framework is supposed to provide a universal solution strategy for this kind of experimental data processing problems in the future.

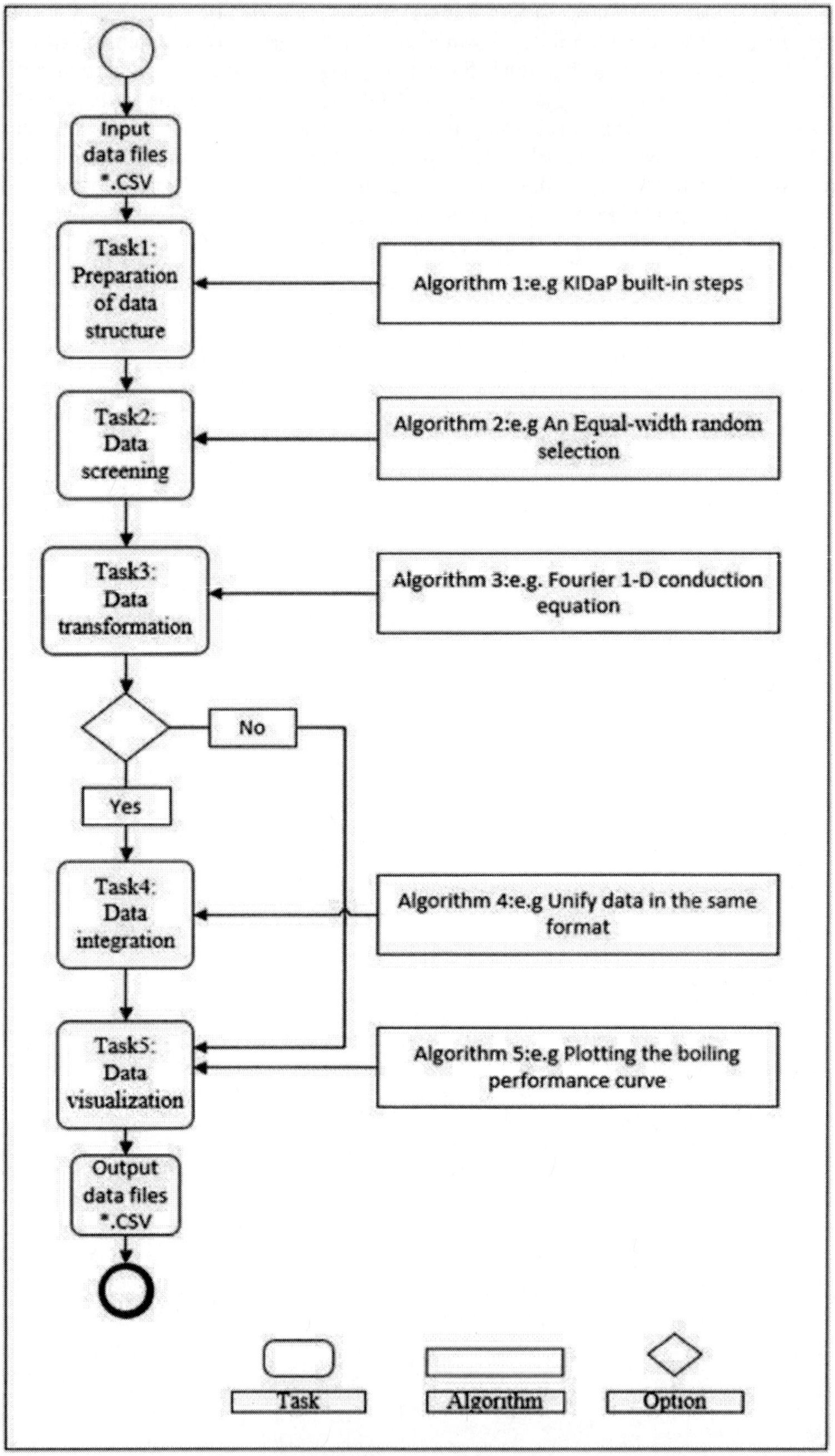

Fig.3. Suggested workflow based on the KIDaP management framework for the example arising from pool boiling experiment

In conclusion, the proposed data preprocessing approach is supposed to provide workflow support and guidance for the experimenters in a more convenient and flexible way. In case of complex work-processes and huge amount of industrial data, the in-depth research of the software framework KIDaP is also supposed to meet the industrial requirement and move one step forward to the development of future intelligent data preprocessing systems, which will be focused on the method of deep learning, the support of the docking of multiple platforms and the basis of the foundation for solving complex sequential as well as parallel problems. In future work, more industrial big data-preprocessing problems arising from different engineering fields are suggested to be tackled by KIDaP based on the proposed management framework. Furthermore, KIDaP will be continuously extended to support intelligent data preprocessing with workflow support so as to satisfy the practical requirement.

Acknowledgments

The authors particularly thank our collaboration partners Dr. Marcus Soemers and Manfred Theissen for providing the open source software KIDaP and the corresponding work-process compendium. Dr. Ya-qiao Wang and Hao-ran Lu are appreciated for giving the measurement data from their pool boiling experiments.

References

[1] G.F., Wang., X.T., Tian., J.H., Geng., et., al. A Process Innovation Knowledge Management Framework and its Application. Advanced Materials Research, 655-657 (2013) 2299-2306.

[2] P., Gaj., A., Malinowski., T., Sauter., et., al. Guest editorial: Distributed Data Processing in Industrial Applications. IEEE Transactions on Industrial Informatics, 11(3) (2015) 737-740.

[3] A., Jashapara. Knowledge Management, An Integrated Approach. Prentice Hall, (2004) 73-108.

[4] E., Žižmond., M., Novak. Controversies of Technology Convergence within The European Union. Industrial Management & Data Systems, 107(5) (2007) 618-635.

[5] S., Natek., D., Lesjak. Improving Knowledge Management by Integrating Hei Process and Data Models. Journal of Computer Information Systems, 53(4) (2015) 81-86.

[6] Z.P., Zhang., S.M., Jasimuddin. Knowledge Market in Organization: Incentive Alignment and IT Support. Industrial Management & Data Systems, 112(7) (2012) 1101-1122.

[7] Y., Wang., S., Yu., T., Xu. A User Requirement Driven Framework for Collaborative Design Knowledge Management. Advanced Engineering Informatics, 33 (2017) 16-28.

[8] D., Mishra., S., Aydin., A., Mishra., et., al. Knowledge Management in Requirement Elicitation: Situational Methods View. Computer Standards & Interfaces, 56 (2018) 49-61.

[9] Y., Heng., M., Soemers., M., Theißen. Wissensbasierte Aufbereitung von Prozessdaten mit Workflow-Support. Mitteldeutsche Mitteilungen, 1 (2012) 8-9.

[10] Y.Q., Wang., J.L., Luo., D.C., Mo., et., al. Wettability Modification to further Enhance the Pool Boiling Performance of the Micro Nano Bi-porous Copper Surface Structure. International Journal of Heat and Mass Transfer, 119 (2018) 333-342.

[11] C., Li., Z., Wang., P.I., Wang., et., al. Nanostructured Copper Interfaces for Enhanced Boiling. Small, 4(8) (2008) 1084-1088.

570790 Data-Driven Predictive Monitoring and Operation Support for Change-over Processes in Biopharmaceutical Drug Product Manufacturing

In biopharmaceutical drug product manufacturing, product-change-over is a crucial but time-consuming process. This step is necessary to establish and maintain a sterile production environment, which is essential to guarantee the product quality. During the change-over process inside the isolator, both the exterior and interior surface of the production machinery require intensive cleaning/disinfection, achieved through hydrogen peroxide decontamination, and clean-in-place (CIP)/sterilization-in-place (SIP), respectively. Numerous sensors are available to constantly monitor the process and provide real-time measurements of process variables, whereby the stored sensor data is only used for backtracking actions. In our work, we exploit the available data to develop a system for predictive monitoring of the process and early failure detection.

The data is initially prepared for exploitation through several steps. First, a dimensionality reduction step through principal component analysis (PCA) is conducted followed by a data cleaning and classification step, where the runs are classified into failed and successful. The boundaries of successful runs are then defined and are further referred to as "Golden Zone." Machine learning algorithms such as Random Forest and k Nearest Neighbor are used to predict the performance of a run following a few minutes of operation. The quality of the prediction is assessed, and the distance to the "Golden Zone" boundaries is measured to guide the decision-making process by the operators in case abnormalities are detected.

The application of this predictive monitoring approach for early failure detection to the historical data of a hydrogen peroxide decontamination process of an industrial filling line at Hoffmann La Roche in Switzerland yielded a 50% reduction in the number of repeated runs due to failures leading to significant time and financial savings.

A complementary root cause analysis is also conducted to deepen the understanding of the complex system where a limited understanding of the causes of failure is provided. Pattern recognition within the analyzed data is used to identify different sources of errors and to recognize the full impact of certain abnormalities or failures (such as leaks) on the system operation and performance.

We further explore opportunities to improve the machine learning algorithms to achieve the earliest possible failure detection points, i.e., to maximize the time distance from the predicted failure to the occurrence of a recorded alarm. This could be especially helpful in the case of hydrogen peroxide decontamination where failures can be detected before the introduction of the hydrogen peroxide to avoid the need to repeat the aeration process which is the most time-consuming step in the procedure.

574765 A Decision Support Tool and Closed-Loop Startup Framework for a Hydrogen Plant

Hydrogen is one of the most extensively used chemical components, with wide range of applications in petroleum and chemical industries. An economic way to commercially produce hydrogen is to utilize steam methane reforming [1]. The process forms an intricate network of material and energy due presence of vast number of material and heat flows. Startup of a hydrogen plant involves implementation of a predefined recipe that consists of a series of events taking place over a finite duration of time. Specifically, it involves continuous actions such as ramping the natural gas fuel to reach a sufficient reformer exit temperature, regulation of nitrogen flow in the reactor tubes, circulation of natural gas feed and steam etc. Further, discrete decisions such as starting up the PSA, recirculating the tail gas from the PSA and ramping up to full capacity in several stages are also part of the startup recipe. Based on the decisions that are taken at various stages in the startup, the startup time can vary significantly. Startup is typically categorized as cold startup and warm startup, based on the initial state of the plant from which the plant is required to be driven to nominal operation. The cold startup will be central to the discussion in this work, where the plant is started from completely shutdown conditions.

Existing startup recipes have been built heuristically from past successful startups, with the possibility of optimizing the startup procedure remaining unexplored. Further, due to the uncertainty associated with the startup duration, the availability of the product cannot be predicted accurately. Motivated by these considerations, in this work we first developed a decision support tool (DST) that can assist the operator during startup to predict the completion of startup. One key recognition that we make is that due to the nature of startup/shutdown phase, they can be treated as a batch like operation with variable lengths, where the objective is to drive the process from one state to the other. Thus, batch subspace identification approach [2] is used for data-driven modeling of the hydrogen plant [3]. Development of a data-driven LTI model, that can handle variable batch duration, paves the way for reliable prediction of the startup evolution and its utilization in a tractable optimization formulation. The DST in real time predicts the timeline of the startup by utilizing the data-driven model and state-trajectories from historical batches. This paper will present the proposed algorithm and results from plant startup.

Secondly, we addressed the problem of synthesis of an optimal startup procedure for the hydrogen production plant. A key requirement for operating procedure synthesis (OPS) is a good model for the startup process. There exist several modeling approaches that are well suited for small scale processes (see [5] for an excellent review) but difficult to implement to large chemical units. Out of these approaches, dynamic simulation based strategies are most prominent. In these approaches, dynamic simulations are used to select the best startup scenarios. In the present work, startup is optimized such that it improves not only the process economics but also its completion time using dynamic optimization. The optimization problem for the present application is an optimal control problem where the objective is to reach the desired terminal state in minimum time, while adhering to various constraints (such as the bounds on reformer exit temperature [6]) and making discrete decisions along the way. To this end, first a high-fidelity test bed model of the entire plant is developed in UniSim, capable of simulating the startup and shutdown phase, with appropriate adaptation of the plant SOP. Note however that such detailed first principles models pose computational challenges when directly embedded in optimization problems, and thus existing SOP synthesis approaches utilize heuristics to determine optimal profile [7-10]. Thus, in this work, several simulated startups are performed to generate the training data. Then, an LTI data-driven model of the process using batch subspace identification [2-4] is identified. The identified data-driven model is subsequently utilized within an optimization framework to synthesize the startup SOP. Through closed-loop startup, we are able to achieve startup in significantly less time. Results from the closed-loop startup and the corresponding optimization problem will be presented.

[1] Kroschwitz, J., Seidel, A., 2006. Kirk-Othmer Encyclopedia of Chemical Technology, No. V. 20 in Encyclopedia of Chemical Technology. Wiley.

[2] Corbett, B., Mhaskar, P. Subspace identification for data-driven modelling and quality control of batch

processes, AIChE Journal 62 (2016) 1581-1601.

[3] Garg, A., Corbett, B., Mhaskar, P., Hu, G., Flores-Cerrillo, J. (2017), High Fidelity Model Development and Subspace Identification of a Hydrogen Plant Startup Dynamics, Computers and Chemical Engineering, 106 183-190.

[4] Garg, A., Mhaskar, P. (2017), Subspace Identification Based Modeling and Control of Batch Particulate Processes, Industrial and Engineering Chemistry Research, 56 (26) 7491-7502.

[5] Batres, R., Soutter, J., Asprey, S.P., Chung, P., 2002. Operating procedure synthesis: science or art. Knowl. Eng. Rev. 17 (3).

[6] Du, M., Mhaskar, P., Zhu, Y., Flores-Cerrillo, J., 2014. Safe-parking of a hydrogen production unit. Ind. Eng. Chem. Res. 53 (19), 8147–8154.

[7] Scenna, N., Ruiz, C., Benz, S., 1998. Dynamic simulation of start-up procedures of reactive distillation columns. Comput. Chem. Eng. 22, S719–S722.

[8] Reepmeyer, F., Repke, J.-U., Wozny, G., 2004. Time optimal start-up strategies for reactive distillation columns. Chem. Eng. Sci. 59 (20), 4339–4347.

[9] Yang, X., Xu, Q., Li, K., Sagar, C.D., 2010. Dynamic simulation and optimization for the start-up operation of an ethylene oxide plant. Ind. Eng. Chem. Res. 49 (9), 4360–4371.

[10] Zhao, Y., Zhang, J., Qiu, T., Zhao, J., Xu, Q., 2014. Flare minimization during start-ups of an integrated cryogenic separation system via dynamic simulation. Ind. Eng. Chem. Res. 53 (4), 1553–1562.

575372 Using Channel State Information for Estimating Moisture Content in Wood Via 5 GHz Wi-Fi

Using Channel State Information for Estimating Moisture Content in Wood via 5 GHz Wi-Fi

Kerul Suthar, Forest Dewberry, Zhihua Jiang, Q. Peter He

Moisture content monitoring is a very important factor in manufacturing for forest industry as well as pulp and paper industry. For a manufacturing process that uses wood as a raw material, it is important to have real-time knowledge about the moisture content of wood so that the process can be optimized and/or controlled to achieve satisfactory product quality. Besides the manufacturing industry, the energy exploitation, storage suitability, and delivery price also depend on the moisture content of wood as well. Both destructive and non-destructive methods have been developed for moisture detection and/or estimation in wood. Destructive methods such as oven drying are accurate, but they also lead to destructive changes in the properties of wood. In addition, they are usually off-line and time consuming and therefore cannot be used for online applications due to the time constraint [1]. On the other hand, several non-destructive techniques that are less time consuming and require less man power have been developed, which include capacitance method, resistance method, microwave method, NIR method, etc. But these techniques are relatively complex and costly to install and operate [2].

To address the above-mentioned limitations of existing techniques, we propose a non-destructive and economic approach based on 5 GHz Wi-Fi, and use Channel State Information (CSI) to predict the moisture content in wood chips. We extract CSI by modifying the open source device drivers for Intel Wi-Fi link 5300 Network Interface Card based on CSITool [3]. CSI contains information about the channel in form of individual data sub-carriers capturing indoor channel features like effect of scattering, fading and power decay with distance. Modern Wi-Fi systems are equipped with Orthogonal Frequency Division Multiplexing (OFDM), dividing the data in multiple orthogonal subcarrier groups which solves the issue of selective frequency fading [4] . CSI has been used for indoor localization, device free sensing including fall detection, activity recognition and heart rate monitoring [5]. In addition, CSI and phase difference data have been successfully used for multi-class moisture detection in wheat [5].

In this work, we collect CSI and phase difference data using IWL5300 NIC by configuring the transmitter and receiver in injection and monitor mode respectively. We use Lenovo ThinkPad systems equipped with Linux based OS 14.02 and kernel version 4.2 due to the version specific selectivity of CSI tool. Both systems are equipped with IWL5300 NIC with modified driver and firmware for data collection. Our work includes CSI data collection, preprocessing, outlier detection, offline training and online testing. The CSI data is collected for wood chips at different moisture levels for offline training of models. The statistics pattern analysis (SPA) framework [6,7] is adapted in this work to build a multivariate statistical model for predicting the moisture content in wood. The most significant difference between SPA based approach and other multivariate statistical methods, such as partial least squares (PLS), is that instead of extracting correlations between the variables and moisture content measurements, we extract correlations between the statistics of the variables and the moisture content measurements. The proposed method is compared with other multivariate statistical and machine learning methods, including PLS, long short-term memory (LSTM) network, etc. We also discuss some challenges of the project and potential solutions to address them.

References:

1. Govett, R., Mace, T., & Bowe, S. (2010). A practical guide for the determination of moisture content of woody biomass. *University of Wisconsin*, 20.

2. John Bogart, Kett US - Learn the six methods for determining moisture: Advantages, Deficiencies and who uses each method.

3. Halperin, D., Hu, W., Sheth, A., & Wetherall, D. (2011). Tool release: Gathering 802.11 n traces

with channel state information. *ACM SIGCOMM Computer Communication Review*, *41*(1), 53-53.

4. Wang, X., Mao, S., Pandey, S., & Agrawal, P. (2014). CA2T: Cooperative antenna arrays technique for pinpoint indoor localization. *Procedia Computer Science*, *34*, 392-399.

5. Yang, W., Wang, X., Cao, S., Wang, H., & Mao, S. (2018, July). Multi-class wheat moisture detection with 5GHz Wi-Fi: A deep LSTM approach. In *2018 27th International Conference on Computer Communication and Networks (ICCCN)*(pp. 1-9). IEEE.

6. Wang and Q. P. He, "Multivariate process monitoring based on statistics pattern analysis," Ind. Eng. Chem. Res., vol. 49, pp. 7858–7869, 2010.

7. P. He and J. Wang, "Statistics Pattern Analysis - A New Process Monitoring Framework and Its Application to Semiconductor Batch Processes," AIChE J., vol. 57, pp. 107–121, 2011.

571273 Identification of Repeating Sequential Alarms in Historical Operation Data of Chemical Plants

The advance of distributed control systems (DCS) in the chemical industry has made it possible to inexpensively and easily install numerous alarms in DCS. While most alarms help operators detect an abnormality and identify its cause, some are unnecessary. A poor alarm system might cause sequential alarms, which are a collection of numerous alarms that almost always occur simultaneously with specific time lags within a short period of time. Sequential alarms reduce the ability of operators to cope with plant abnormalities because critical alarms are buried under many unnecessary ones. It is very important to identify sequential alarms in a plant operation data for safe plant operations.

Nishiguchi and Takai (2010) proposed an identifying method of sequential alarms in plant operation data using event correlation analysis. Event correlation analysis is a method of extracting knowledge that detects statistical similarities between discrete events. The method can identify sequential alarms from the operation data of chemical plant, but it occasionally failed to detect similarities between two physically related sequential alarms when deletions, substitutions, and/or transpositions occurred in the alarm sequence. A modified Smith-Waterman algorithm was proposed to calculate similarities of alarm flood sequences in plant operation data (Cheng *et al.*, 2013). Experts can make a thoroughly analysis, such as root cause, based on the clustered patterns of alarm floods. However it is not the method that directly finds out sequential alarms hidden in the plant operation data. Wang and Noda (2017) proposed an identification method of sequential alarms by applying the dot matrix method to a plant operation data. The dot matrix method is one of sequence alignment methods for identifying similar regions in DNA or RNA, which may be a consequence of functional, structural, or evolutionary relationships between the sequences (Mount, 2004). The method can identify sequential alarms from the operation data of chemical plant, but it occasionally failed to detect similarities between two physically related sequential alarms because the time information when alarms occurred is not for evaluating similarities between them.

In this research, we propose a new identification method of repeating sequential alarms in historical operation data of a chemical plants. In the method, time stamped alarm data is converted into a set of windows containing adjacent alarms. All combinations of windows are compared, and repeating similar windows are identified on the basis of similarities between them. The alarms in each repeating window comprise a sequential alarm. Application of this method to simulated operation data for an azeotropic distillation column demonstrated that it can identify sequential alarms in noisy plant-operation data. By classifying such alarms into small numbers of sequential alarms, an engineer is able to effectively reduce the number of unnecessary alarms.

References

1. Cheng, Y., Izadi, I., and Chen, T., Pattern matching of alarm flood sequences by a modified Smith-Waterman algorithm, *Chemical Engineering Research and Design*, 91, 1085-1094 (2013)
2. Nishiguchi, J., and Takai, T., IPL2 and 3 Performance Improvement Method for Process Safety using Event Correlation Analysis, *Computers & Chemical Engineering*, 34(12), 2007-2013 (2010)
3. Wang, Z., and Noda, M., Identification of Repeated Sequential Alarms in Noisy Plant Operation Data Using Dot Matrix Method with Sliding Window, *Journal of Chemical Engineering of Japan*, 50(6), 445-449 (2017)
4. Mount, D., W., Bioinformatics Sequence and Genome Analysis Second Edition, Cold Spring Harbor Laboratory Press, Cold Spring Harbor, New York (2004)

573955 Big Data for Enhanced Reduced Order Modeling: Application to Hydraulic Fracturing

Many chemical processes undergo non-linear dynamics which are usually described by high-dimensional high-fidelity models whose complexities limit their usage in the design of model-based feedback control systems. These hindrances have paved the way for the use of reduced-order models (ROMs) in the design of controllers. Among the plethora of available techniques, Dynamic Mode Decomposition (DMD) [1], has gained a lot of attention as it accurately captures coherent structures that are dynamically relevant and contribute towards the long term dynamics of the system [2]. The primary assumption of the DMD technique is that the non-linear dynamics of a complex system can be represented in a linear form. But the dynamics of many chemical processes are influenced by the external inputs applied on them and therefore, Dynamic Mode Decomposition with Control (DMDc) technique [3] was developed to include the effect of the external inputs on the dynamics of the system. However, for a highly non-linear chemical process, a global linear representation may not accurately capture its dynamics considering that DMDc has very limited degrees of freedom. To overcome this limitation of DMDc, Local Dynamic Mode Decomposition with Control (Local DMDc) technique [4] was developed which utilizes unsupervised learning to cluster the sequential data, use DMDc to build local ROMs which individually capture the local dynamics accurately and together represent the dynamical system.

However, the domain of attraction (DOA) of Local DMDc is narrow; meaning, it will be less accurate when used for prediction purposes under conditions outside the training data. In this work, we overcome the problem of limited DOA of Local DMDc by using multiple training data sets obtained under distinct training conditions. Specifically, training data is divided into clusters such that each cluster represents certain unique, local dynamics of the system and a DMDc-based local ROM is built for each cluster. Also, a unique identity is built for each cluster by transforming the data through normalization, applying weights, and Principal Component Analysis (PCA). The identity for each cluster is the average of the PCA scores of all the data points it contains. During prediction, given a state and input, we utilize the k-nearest neighbors (kNN) technique to select the appropriate local ROM in order to calculate the future state of the system. We demonstrate the performance of our proposed algorithm by applying it to hydraulic fracturing, a highly non-linear and complex process represented by a system of nonlinear highly-coupled PDEs with time-dependent spatial domain [5]. We utilize its high-fidelity model to generate training data obtained by using distinct training inputs, build multiple local ROMs based on DMDc, and utilize kNN for the purposes of model selection. We demonstrate the enlarged DOA through increased applicability of the LDMDc models obtained using the proposed algorithm for a wide range of operating conditions

Literature cited:

[1] Schmid, P.J., 2010. Dynamic mode decomposition of numerical and experimental data. J. Fluid Mech. 656, 5–28.

[2] Ghommem, M., Carlo, V. M., Efendiev, Y., 2014. Mode decomposition methods for flows in high-contrast porous media. A Global approach. J. Comput. Phys. 257, 400-413.

[3] Proctor, J. L., Brunton, S. L., Kutz, J. N., 2016. Dynamic mode decomposition with control. SIAM J. Appl. Dyn. Syst. 15, 142-161.

[4] Narasingam, A., Kwon, J. S., 2017. Development of local dynamic mode decomposition with control: Application to model predictive control of hydraulic fracturing. Comput. Chem. Eng. 106, 501-511.

[5] Economides, M.J., Nolte, K.G., 2000. Reservoir stimulation. Chichester, Wiley.

564569 A Clustering Decomposition Algorithm for Energy Storage Design & Operation

Intermittent availability of solar and wind energy is a challenge for renewable power systems because they are often asynchronous with consumer demand. Increasing renewable penetration requires energy storage technologies to align varying supply with demand. Moreover, energy storage options give greater control and flexibility to grid operators by making solar and wind more dispatchable power sources.

Optimization-based design & scheduling models aim to minimize the power and storage capacities of renewable power systems to lower capital and operational costs [1]. Hourly time discretization in the schedule is often used to capture solar and wind dynamics, keep track of storage inventory levels, and model time-dependent operational decisions. As a result, large time horizons are characteristic of these problems and significantly increase the computational burden of solving them [2, 3]. In addition, multiple time series data coming from resource availability, demand loads, and prices grow the complexity as well.

In this work, a decomposition algorithm based on agglomerative hierarchical clustering (AHC) is proposed to alleviate the computational burden, where the optimization is performed over representative time periods. A key advantage for AHC compared to the popular K-means clustering approach is that the clusters maintain time chronology, which is important for analyzing inter-period energy storage [4]. An example case study on dense energy carriers for energy storage, performed in collaboration with Shell, is presented to demonstrate the algorithm's applicability.

[1] Zhang, Qi, Ignacio E. Grossmann, Clara F. Heuberger, Arul Sundaramoorthy, and Jose M. Pinto. "Air separation with cryogenic energy storage: optimal scheduling considering electric energy and reserve markets." *AIChE Journal* 61, no. 5 (2015): 1547-1558.

[2] Harjunkoski, Iiro, Christos T. Maravelias, Peter Bongers, Pedro M. Castro, Sebastian Engell, Ignacio E. Grossmann, John Hooker, Carlos Méndez, Guido Sand, and John Wassick. "Scope for industrial applications of production scheduling models and solution methods." *Computers & Chemical Engineering* 62 (2014): 161-193.

[3] Castro, Pedro M., Ignacio E. Grossmann, and Qi Zhang. "Expanding scope and computational challenges in process scheduling." *Computers & Chemical Engineering* 114 (2018): 14-42.

[4] Pineda, Salvador, and Juan M. Morales. "Chronological time-period clustering for optimal capacity expansion planning with storage." *IEEE Transactions on Power Systems* 33, no. 6 (2018): 7162-7170.

567770 Process Intensification of the Alkoxylation of Fatty Amines

Process Intensification of the Alkoxylation of Fatty Amines

Pia Müller[1*], Wyatt Winkenwerder[2], John van der Schaaf[1]

1 Department of Chemical Engineering and Chemistry, Eindhoven University of Technology, P.O. Box 513, 5600MB, the Netherlands; 2 Nouryon, Brewster, New York 10509, United States

*Corresponding author: p.muller@tue.nl

j.vanderschaaf@tue.nl

wyatt.winkenwerder@nouryon.com

1. Introduction

Alkoxylated fatty amines are widely used as non-ionic surfactants, leading to a steady increase in industrial demand over the past six decades. These compounds are key components in products of numerous industrial branches, e.g. oil drilling, paper manufacturing, textile industry, as well as fiber manufacturing, paints, dyes and plastics [1]. These compounds are typically produced by reacting fatty amines with ethylene oxide (EO) or propylene oxide (PO) in large fed-batch reactors, often reaching the size of 30 m^3. The process is intentionally de-intensified, due to its strongly exothermic nature ($\Delta_R H$ = -120 kJ/mole of EO) and the safety risks associated with EO. Therefore, the epoxide is slowly added to the fatty amine to keep the exothermic reaction under control. In the past, attempts have been made to intensify the reaction by using venture-loop or spray-tower reactors [2], [3]. However, due to the required active mixing, these processes necessitate a power input higher than the typical set-up, and fail to provide significantly improved product output. The aim of our work is to transfer this industrially important reaction to using modern technology to a 21st century process. One possible option to be is shown in Figure 1.

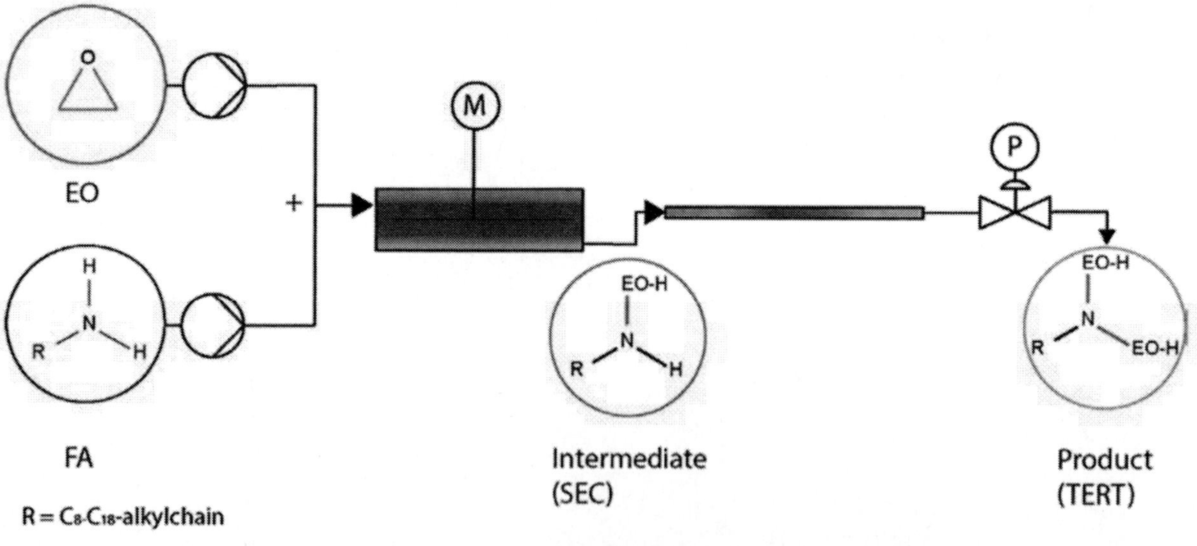

Figure 1: Schematic of possible, intensified reaction setup. In total two mole EO (or other epoxides) react with the fatty amines to

form a di-alkoxylated amine. The process includes a CSTR in series with a plug flow reactor.

This study focuses on gaining detailed kinetic insights and their application to the development of continuous flow systems to improve the reaction efficiency whilst maintaining safety standards.

2. Methods

To achieve our aim a kinetic study was necessary. In a first step we used model compounds such as butylene oxide (BO) or propylene oxide (PO) with octylamine (OA) in a lab scale reactor. A temperature window of 120 °C to 160 °C was explored by analyzing the mixture with GC, NMR and inline FTIR. The process imitated the industrial process by adding the epoxides in a lower molar ratio stepwise to the amine.

In a second steps the collected data were then applied to the design of a flow set-up utilizing microchannels as perfect plug-flow reactors. A comparable flow system was also employed to determine the kinetics of ethylene oxide (EO) with OA over a temperature range between 120 and 170 °C. For safety reasons and accurate contact times it was important to work in a fully liquid system. The system was therefore pressurized by a back pressure regulator to 12 bar (PO and BO) or 60 bar (EO). Such pressure for PO was chosen after a large negative deviation from ideal mixing (8 bar measured at 150 °C to 22 bar in mixture) behavior was observed and a NRTL model was designed. For EO a high tolerance of the theoretical vapor pressure was chosen.

Finally, the reactions with PO and OA was transferred to different set-ups of potential industrial interest. Options like spinning disc reactors or other CSTR types were tested, as well as the recycling of an intermediate mixture

3. Results and discussion

For the reaction of BO and PO with octylamine experimental data were collected in semi-batch and flow:

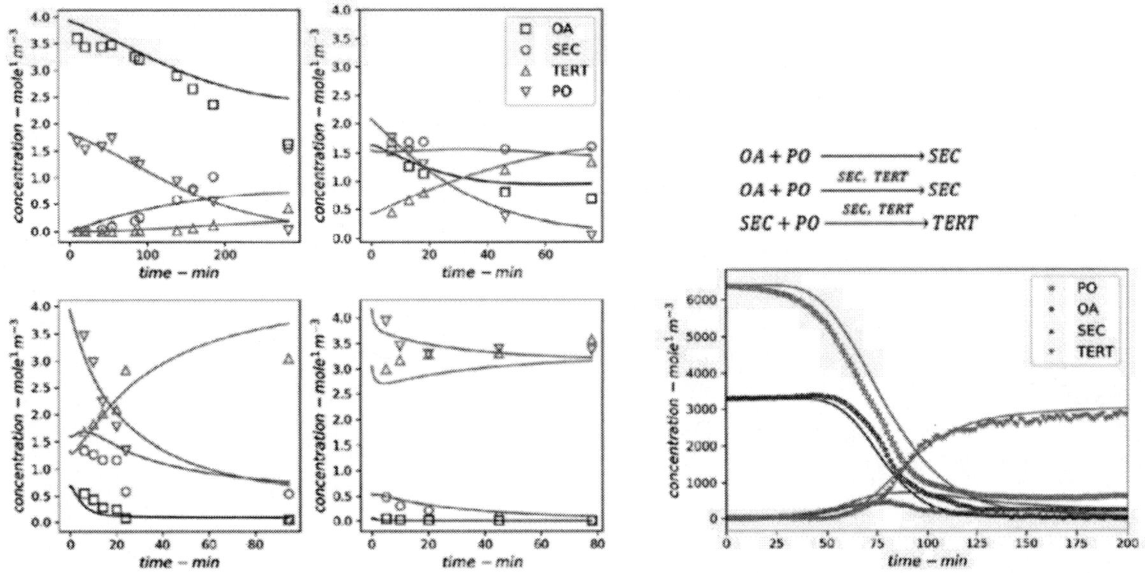

Figure 2: Experimental data in fed-batch (left) was collected over four addition steps of PO of each 0.5 mole per mole OA. In flow (right) the molar ratio initially was 2 mole PO per mole. Both experiments were performed at 130 °C. An auto-catalytic model (upper right) has been used to simulate both experiments.

The reaction is initially very slow in both batch and flow. The reaction accelerates after a certain amount of product is formed, which is typical for auto-catalytic reactions. The advantages of going from fed-batch to flow are clearly visible in Figure 2. What is impossible to achieve in batch, i.e. high conversion in only one step, is feasible in flow. Through the use of microchannels the volume could be decreased from 100 to 7 ml. Thanks to the intensified mixing the reaction time for achieving full conversion could also be decreased by a factor of 1.5-3.0 depending on the epoxide and conditions. Furthermore, the continuous set-up is superior due to the higher saturation of the amine with epoxides achieved under pressure. The overall reaction for PO was 1.5 times as fast as for BO. It is expected for EO an increase in rate by at least a factor 2 compared to PO. The use of a CSTR will lead to an additional decrease of reaction time, due to a better back mixing, which would enhance the inherent auto-catalytic effect of the reaction.

4. Conclusions

Process intensification for the alkoxylation of fatty amines was achieved with a test flow system. A rate increase by a factor 1.5-3.0 could be reached for the reaction of propylene and butylene oxides with the model substrate octylamine. Using a continuous microchannel flow system, the intrinsic safety risks associated with the process could be overcome. A kinetic model that fits the experimental data for the reactions of BO, PO, and EO with amines was found. Using this model, different process configuration options are easily compared for performance.

References

[1] T. M. Schmitt, *Analysis of Surfactants, Second Edition*. CRC Press, 2001.

[2] L. L. van Dierendonck, J. Zahradnik, and V. Linek, "Loop Venturi Reactor A Feasible Alternative to Stirred Tank Reactors?," *Ind. Eng. Chem. Res.*, vol. 37, no. 3, pp. 734–738, 1998.

[3] M. Di Serio, R. Tesser, and E. Santacesaria, "Comparison of Different Reactor Types Used in the Manufacture of Ethoxylated, Propoxylated Products," *Ind. Eng. Chem. Res.*, vol. 44, no. 25, pp. 9482–9489, Dezember 2005.

568871 Process Intensification of Propane Dehydrogenation Using Microporous Silica Membranes

Propylene is an important intermediate chemical for production of petrochemicals such as polypropylene and acrylonitrile. Propylene and other light olefins are mainly produced by light hydrocarbon fractions such as steam cracking and fluid catalytic cracking. The equilibrium limit exists in traditional packed bed reactors (PBR), thus packed bed membrane reactors (PBMR) became an important subject to explore because they can combine chemical reaction and separation in one step. Due to selective removal of one of the products (H_2), PBMR helps in shifting the equilibrium towards the forward reaction, and can overcome the equilibrium limitations in PBR. In this work, we developed microporous silica membranes modified with atomic layer deposition (ALD) for effective H_2 separation from hydrocarbons. The ALD-modified silica membrane was used in PBMR for PDH reaction in combination with Pt/Al_2O_3 catalyst. Moreover, one-dimensional plug-flow reactor (PFR) model was established to investigate the effect of membrane properties and operating conditions on PDH reaction systems.

570440 Scalable Thin-Layer Membrane Reactor for Heterogeneous and Homogeneous Catalytic Gas–Liquid Reactions

Catalytic gas-liquid reactions have potential as environmentally benign methods for organic synthesis, particularly hydrogenation and oxidation reactions. However, safety and scalability are concerns in the application of gas-liquid reactions. In this work, we develop and demonstrate a scalable, sustainable, and safe thin-layer membrane reactor for heterogeneous Pd-catalyzed hydrogenations and homogenous Cu(I)/TEMPO alcohol oxidations. The implementation of a Teflon amorphous fluoroplastic (AF) membrane and porous carbon cloth in the membrane reactor provides sufficient gas-liquid mass transfer to afford superior performance compared to conventional packed-bed or trickle-bed reactors. The membrane separates the gas from the liquid, which avoids the formation of explosive mixtures for oxygenation reactions and simplifies the two-phase hydrodynamics to facilitate scale-up by stacking modules, while significantly reducing gas consumption. In addition, 3-dimensional simulations deliver insights into the mass transfer and hydrodynamic behavior to inform optimal membrane reactor design and operation.

577756 Plug Flow Membrane Reactor for Esterification of Acetic Acid with Mor-Type Zeolite Membrane

While esterification is important for chemical production, the attainable level of conversion is always thermodynamically limited. Thus, the product of esterification is a mixture of organic acid, alcohol, ester, and water. Dehydration membrane reactor has been drawn attention as a novel energy-saving process for esterification for recent decades. When a dehydration membrane is adopted to remove water from the reaction system, the attainable level of conversion will be improved beyond that limited by thermodynamics in conventional reactor. Consequently, one can expect that energy consumption for purification process after reaction is reduced as well as the improvement of one-path yield of products.

We previously reported MOR-type zeolite membrane for dehydration and pointed out that our MOR membrane had high chemical resistance for organic acid.[1] In this study, we prepared applied MOR membrane to a plug flow reactor for the esterification of acetic acid.

MOR membranes were synthesized by a seed-assisted crystallization method on the outer surface of porous tubular a-alumina support (Noritake, i.d. = 7 mm, o.d. = 10 mm) of which the average pore size of the top layer was ca. 150 nm. The outer surface of support was seeded by a dip-coating method using a colloidal suspension of MOR seed crystals.

The seeded support was placed vertically in a Teflon-lined stainless steel autoclave and filled with a reaction mixture. The reaction mixture had a molar composition of $10Na_2O:0.15Al_2O_3:36SiO_2:960H_2O$ and was prepared by mixing colloidal silica (ST-S, Nissan Chemical Ind. Ltd.), sodium aluminate (Kanto Chemical Co. Inc.), sodium hydroxide (>97 wt%, Kanto Chemical Co. Inc.), and deionized water. The reaction mixture was aged at 323 K for 4 h prior to use. Hydrothermal crystallization was carried out at 453 K for 7 h in an autoclave. After the crystallization, the autoclave was quenched with water. The synthesized membrane was washed in boiling water and dried over night at 383 K.

Dehydration performance of MOR membrane prepared was evaluated by the separation test for an equimolar mixture of water and acetic acid in a pervaporation mode at 343 K. The permeation side, the inner side of tubular membrane, was evacuated by using a vacuum pump. The permeate was condensed by a cold trap and was analysed for its composition by using GC-TCD.

The esterification of acetic acid was carried out in a plug flow membrane reactor as follows. The equimolar mixture of acetic acid and ethanol was fed to the outer surface of tubular membrane. Amberlyst-15 was packed on the outer side of membrane as catalyst. Temperature and pressure were adjusted at 403 K and 1.0 MPa, respectively. The permeate was analyzed for its composition. The yields of ethyl acetate produced in the plug flow reactor with and without MOR membrane were compared.

As a result of separation test for acetic acid and water, MOR membrane showed great dehydration performance. The separation factor for water/acetic acid was above 139 with its permeance of 1.65×10^{-7} mol m^{-2} s^{-1} Pa^{-1}. Separation and permeation performance were stable at least up to 8 h, suggesting that MOR zeolite is one of the feasible membrane materials for dehydration under acidic conditions.

Whereas the yield of ethyl acetate in the plug flow membrane reactor without MOR membrane was thermodynamically limited at 66 %, its yield was improved up to 72 % with the membrane in the reactor tested. Selective removal of water by MOR membrane contributed to the enhancement of ethyl acetate yield.

(1) G. Li, E. Kikuchi, M. Matsukata, The control of phase and orientation in zeolite membranes by the secondary growth method, *Microporous Mesoporous Mater.* 62 (2003) 211-220

Acknowledgment

This work was partially supported by NEDO (New Energy and Industrial Technology Development Organization), Japan.

563345 Continuous-Flow Synthesis of the Azo Pigment Yellow 14 Using a Three-Stream Feeding Micromixer

Abstract:

Organic coloured pigment is one of the most important chemical fines. The demand for organic coloured pigments continuous to grow with the growing use in the field of colour printing、cosmetics、coatings、plastics and so on. Among the organic coloured pigments, the most well-known class is the azo pigment. One example is azo pigment yellow 14, which is a typical bisazo pigment. The yellow pigment is usually the last component of the printing process[1] and therefore, they often need to have some special properties, such as a high transparency、a high strength of colour and a high tinctorial power[1]. The colour properties of azo pigments are connected with the particle size and the particle size distribution. They are affected by the production process [1].

In industry, the azo pigments are often synthesized by batch process using stirred tanks with volumes varied from 20 m^3 to 80 m^3[1]. However, batch processes are possible to lead to a relatively board size and a low conversion of the coupling component, which seriously affects the quality of the azo pigments. For the synthesis process of azo pigment yellow 14, it includes double diazotization and double coupling reaction, which need an accurate regulation of the pH during reaction process. Therefore, a better mixing effect and a more precise adjustment of synthesis process are needed. In last 20 years, the microreactor technology has been widely used in both industry and academia[2]. Compared with traditional batch vessels, the microreactor technology has many advantages, such as better mass and thermal transfer, more accurate control parameters[3].

The first successful synthesis of azo pigments in microfluidic devices was reported by Wille et al.[4] The azo pigments showed improvements in color properties compared with batch processes. Then, the synthesis of Pigment Yellow 12 in a micromixer apparatus was developed by Pennemann et al.[1] Using a Micromixer-based process, they improved the properties of the azo pigment yellow 12. However, the azo pigments synthesized above were not sensitive to the changing of the pH during reaction process compared with the azo pigments yellow 14.

In this work, a three-stream feeding process for the synthesis of azo pigment was considered to control the pH of the coupling reaction process accurately. The acetic acid/sodium acetate buffer solution was pumped into the micromixer as a separate stream which was different from the two-stream feeding process and batch process. The pH of the reaction process was not necessary to be monitored at any time if the pH of the buffer solution was controlled. Furthermore, a three inlets microreactor for easy scale-up study was designed to carry out the coupling reaction. Then, the effects of pH of the buffer solution、reactants flow rates and reaction temperature on the purity and the particle size distribution of the azo pigments were discussed. Finally, the hue and transparency of the products compared with pigments standard (commercial products, SHENYANG RESEARCH INSTITUTE OF CHEMICAL INOUSTRY) were also studied. The results show that under optimal conditions, a higher purity (94.54%) of the pigment can be obtained using three-stream feeding process than that using two-stream feeding process (83.34%) and a monomodal particle distribution can be obtained compared the batch process getting a bimodal distribution. According to the coordinates of CIELab colour space (L*, a*, b*) evaluated for azo pigments, it illustrated that the red hue and the blue hue of the azo pigment products were larger compared with pigment yellow 14 standard. The lightness L* of the pigments synthesized by three-stream feeding process was larger than that by two-stream feeding process. The transparency was increased by 98% compared to a commercially available yellow 14. Overall, the micromixer-based three-stream feeding process led to a significant improvement of the azo pigment yellow 14 process, which had a good industrial application value.

References:

[1]Pennemann, S. Forster, Kinkel, Improvement of Dye Properties of theAzo Pigment Yellow 12 Using a Micromixer-Based Process. Org. Process Res. Dev. 9 (2005) 188-192.

[2]J.S. Zhang, K. Wang, Y. C. Lu, G. S. Luo, Characterization and modeling of micromixing performance in micropore dispersion reactors. Chem. Eng. Prog. 49 (2010) 740-747.

[3]Z. Q. Yu, X. Ye, Q. L. Xu, X. X. Xie, H. Dong, W. Su, A fully continuous-flow process for the synthesis of p-Cresol:Impurity Analysis and process optimization. Org. Process Res. Dev. 21(2017) 1644-1652.

[4]C. Wille, V.Autze, H. Kim, U. Nickel, S. Oberbeck, T. Schwalbe, L. Unverdorben, In Proceedings of the 6th International Conference on Microreaction Technology, IMRET 6, March 11-14,2002, New Orleans, LA, U.S.A.; pp 7-17.

AUTHOR INDEX

AUTHOR INDEX

AUTHOR INDEX

AUTHOR INDEX

AUTHOR INDEX

AUTHOR INDEX

AIChE
120 Wall Street, FL 23
New York, NY 10005-4020

ISBN 978-1-7138-0555-7

CURRAN ASSOCIATES INC.
proceedings
.com